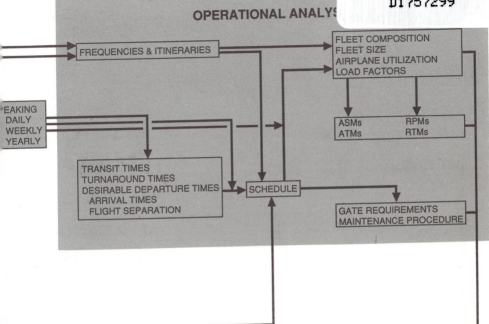

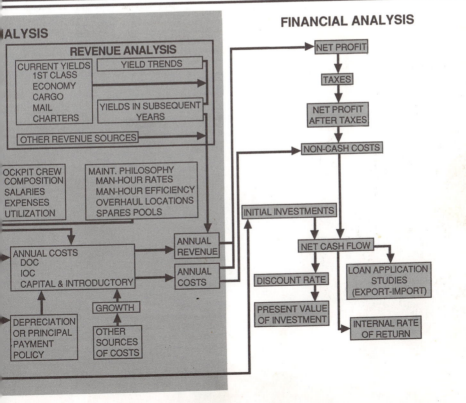

Academic Services
Library Services

Airline Management

Airline Management

Charles F. Banfe

Lecturer
Department of Industrial Engineering
and Engineering Management
Stanford University

PRENTICE HALL
Englewood Cliffs, New Jersey 07632

Library of Congress Cataloging-in-Publication Data

Banfe, Charles.
    Airline management / by Charles Banfe.
        p.   cm.
    Includes index.
    ISBN 0-13-019183-3
    1. Airlines--Management.   I. Title.
HE9781.B36   1992
387.7'068--dc20   91-14221
                                                          CIP

Acquistions editor: DOUG HUMPHREY
Editorial/production supervision and
  interior design: RICHARD DELORENZO
Copy editor: JAMES TULLY
Cover design: LUNDGREN GRAPHICS
Cover photo: OLAF SÖÖT/TONY STONE WORLDWIDE
Prepress buyer: LINDA BEHRENS
Manufacturing buyer: DAVID DICKEY

© 1992 by Prentice-Hall, Inc.
A Simon & Schuster Company
Englewood Cliffs, New Jersey 07632

Printed in the United States of America

10 9 8 7 6 5 4 3 2 1

ISBN 0-13-019183-3

Prentice-Hall International (UK) Limited, *London*
Prentice-Hall of Australia Pty. Limited, *Sydney*
Prentice-Hall Canada Inc., *Toronto*
Prentice-Hall Hispanoamericana, S.A., *Mexico*
Prentice-Hall of India Private Limited, *New Delhi*
Prentice-Hall of Japan, Inc., *Tokyo*
Simon & Schuster Asia Pte. Ltd., *Singapore*
Editora Prentice-Hall do Brasil, Ltda., *Rio de Janeiro*

TO MY JUNE

Contents

Preface

This is a "how it is done" book. It will provide readers with information on an industry in dynamic motion that involves many separate, yet related areas of understanding.

For those who do not fully understand the air transportation industry, this book will be a general overview from a management perspective. For those in specialized sectors of the industry, it will attempt to carefully explain other functions in air transportation. Whatever the area of interest, this book hopes to cover the industry at a comprehensible level without the need of a highly technical background.

The objective of this book is to provide data and information which will allow readers to identify differing management structures, techniques, and problems within the air transportation industry, both in the United States and abroad.

In the United States, commercial air transportation is relatively straightforward. The mechanics of the business and the way it is conducted are common to all carriers. Because of deregulation, varying histories, resource availabilities, unusual emphases by Chief Executive Officers, airline firms have followed surprisingly divergent paths in identifying markets, establishing opportunities, and constraints in the environment.

This also holds true to a slightly lesser degree for the international air transportation industry. However, while U.S. carriers are all privately held, about 80% of the airlines abroad are government instruments or controlled and protected in the national interest. Thus, their regulatory behavior may be different. Foreign governments may establish revenue and profit reporting procedures which may not allow the same public disclosure of information as is required of U.S. carriers under Security Exchange Commission law. In the international arena, it is not uncommon for governments to regulate capacity, frequency, pricing, reporting, cost allocation, goals, and actions. To the extent that this is not parallel to the practices under the law in the United States, results may not bear comparison.

Therefore, though the U.S. and international air transportation industries are similar in many basic respects, they are, in fact, clearly differentiated by the level of government participation.

The book begins with a history and background of the industry. This is followed by a description which may seem awkward but is necessary for the rational evolution of air transportation and management. Each discipline within an airline is discussed in general terms. Certain important challenges for air transport management will emerge in the course of discussion.

This book lays no claim to be incontrovertible, original, or profound. After listening to about half a hundred presidents describe the air transportation industry and how it functions, one recognizes that value judgments and personal styles play a pivotal role. It is a dynamic industry, fosters controversy and dissension, but there is a discernible thread passing through the book which is common to successful management and executive practices. The objective is to be informative, to provide points of view and tools for those who want to understand what makes johnny run.

It is, to my knowledge, the first attempt at an organized presentation of airline management, how it functions behind the corporate doors, and a halting stride toward understanding the rationale and forces acting on those involved in airlines and those who make the decisions.

If there could be a guarantee on "how it is done" it would be to describe airline management, not as a conventional textbook presentation, but the synthesis and annealing of real-world air transportation management as described by senior airline officers.

As a last comment, the more than one hundred senior officer distinguished guest lecturers were carefully selected and associated with successful carriers. The belief was that more could be learned from success than from failure. It was found that there is an underlying constancy of management beliefs and practices in successful air carriers.

Not all of the material contained here is of my derivation. Much of it has been the result of research and may resemble material done by others. However, this book represents a concerted effort to gather fractionalized material and put it into one cohesive book for clearer understanding.

Acknowledgments

Two persons made this book eminently more useful than it otherwise might have been—Bill Crown and Jahan Alamzad.

Bill took the original manuscript, much too long, too redundant, too rambling and blasted away to make it very readable. Jahan fine-tuned the book and did a great job!

Stanford University was gracious in its support and, in particular, the head of my department, Industrial Engineering and Engineering Management. Dr. Warren Hausman has always been a champion and a friend.

Stanford University supported the writing of this book but did not influence its content nor does it reflect an academic philosophy. It is mine, thirty-five years of being in the industry in many capacities, and being a student of management all over the world.

Bill Crown is Director–Domestic Passenger Pricing, American Airlines. Jahan Alamzad is Lecturer of Airline Management at San José State University, and Principal, Management Science Group.

PART 1

Development of the Air Transport Industry

Chapter 1

Introduction
and History

Courtesy of Boeing Commercial Airplanes; Boeing 747-400

THE BEGINNING

Man's progress has always been linked to his ability to travel. The earliest man, who could travel only by foot, knew little or nothing about his world outside a 10-mile radius. Therefore, his tribe represented all mankind to him. As animals were domesticated and used for transportation and, more dramatically, when bodies of water were made navigable by the invention of the dugout canoe, man's boundaries expanded and progress was increased markedly. For the next 4000 years the advantages of travel caused civilization to advance most rapidly near bodies of water.

Although early legends, history, and literature demonstrate interest in the possibility of human flight, the conquest of the air did not occur until the modern era. Early on, the Chinese had developed five devices that utilized properties of flight; the arrow, the windmill, the boomerang, the rocket, and the sail. It was not until 1783, however, that the first successful experiments in human flight took place—outside Paris in a hot-air balloon. It was nearly another hundred years, in 1874, before Count von Zeppelin, who experimented with a rigid airship powered by an engine and a propeller, took the next major step. Finally, on December 17, 1903, the Wright brothers introduced the Golden Age of aviation when they completed the world's first powered, sustained, and controlled airplane flights at Kitty Hawk, N.C. Theirs was the first machine that could bank, turn, and circle with comparative ease. Although that first flight of 120 feet could be duplicated today inside a Boeing 747 (and with room to spare), it nonetheless marked the beginning of controlled heavier-than-air flight.

All the while, the commercial potential of aircraft was attracting attention. In 1908, Charles Furnas, an American, became the first commercial passenger ever to fly. In late 1909, Count von Zeppelin's pencil-shaped craft brought Germany the distinction of providing the world's first scheduled passenger air service, which began at Lake Constance. Between 1910 and 1914, five Zeppelins carried 35,000 passengers a total distance of 195,500 kilometers between Lake Constance and Berlin and other German cities without a single fatality. On the equipment side, France was the first nation to establish an airplane industry, and it rapidly emerged as the acknowledged center of world aviation and maintained its supremacy for generations.

France's leadership in aviation did not go unchallenged, however. Then, as now, the power plant held the key to progress in flight. More powerful engines were designed, propellers were shifted from the pusher position in the rear to the front of the aircraft, the fuselage was enclosed, landing wheels were added, control surfaces were made more effective, and perhaps more important, box-kite structures were modified to slender airplanes capable of airspeeds of up to 300 kilometers per hour (hereafter km/h). Notable in that brief but dramatic period between 1910 and 1914, Igor Sikorsky pioneered the development of large multiengine transports in Russia, and Hugo Junkers produced the first all-metal, fully cantilevered wing monoplane. As aircraft improved, the use of air transportation spread rapidly.

While the Europeans were concentrating on developing passenger air service, the United States focused on carrying airmail. In 1910 the first airmail bill was introduced in Congress. Although defeated, the bill marked the beginning of commercial aviation in America. In the following year the first 4-pound sack of mail was flown by the United States Post Office Department in a French Bleriot Queen monoplane. By 1912, some 31 mail flights had taken place and the service continued to expand.

In 1914, P.E. Fansler inaugurated the first regular, domestic air passenger service across Tampa Bay in another French aircraft, the Benoist flying boat. The 35-mile trip took 23 minutes flying time and carried one passenger each way for a $5 fare.

WORLD WAR I AND LATER

When war broke out in Europe in August 1914, it was clear that air transportation would be an important logistical tool. When the hostilities began, there were only 200 Allied aircraft and about 180 German air machines on the Western Front. The emphasis placed on the developments and improvements in both aircraft design and manufacture accelerated progress in aviation. For the first time, organized research, with strong government support, was centered on aircraft and engine design. Under this intense cultivation, the airplane grew in power, speed, and dependability.

After the Armistice, the transition to civil aviation and commercial air transportation moved slowly and erratically. The year 1919 marked the start of commercial air transportation in fixed-wing aircraft. Bolder than their American counterparts, six major European airlines can trace their beginnings to that same year. The International Air Transport Association (IATA) was formed and the first scheduled London-to-Paris service was inaugurated. Three British airlines introduced service that was to last for two years. In Germany, air service began among the cities of Berlin, Leipzig, and Weimar. France also quickly set about the development of air services after the war. Regular scheduled international passenger service began in a Farman Goliath with 11 military passengers. Numerous other European countries began air transport operations including Denmark, The Netherlands, and Switzerland.

In Europe and many of the industrialized countries of the world, with the exception of the United States, national airlines began to emerge, often under private ownership but always with strong government support. France, Germany, and The Netherlands took the lead, soon to be followed by Great Britain, Switzerland, and other European nations. Why Europe developed air transportation so vigorously is somewhat of a paradox. Distances on the Continent were relatively short compared with those in the United States and elsewhere. The primary function of air transport was to reduce travel time, particularly those of great distances, and it would seem that this need was not as pronounced in Europe. Nevertheless, it is sufficient to note that America entered air transportation by opening up a nationwide airmail service while the Europeans perceived the airplane as a carrier of passengers. Even with the finest railroad network in the world, Europe soon became interlaced with air routes.

The United States was alone in its preoccupation with mail, showing little concern for passengers. In the Western Hemisphere, the airlines of Colombia and Australia can trace their beginnings to 1920. Colombia was the first nation to attempt to operate commercial air services in the tropics. Australia's initial air service was inaugurated between Charlesville and Conclurry in an Armstrong Whitworth FKS. In the following year, Belgium, Spain, and New Zealand had air service. In 1922, regular flights were launched in Argentina, Japan, Canada, and Poland.

The decade of the 1920s was dominated by trimotor aircraft built by Germany (Junkers) and The Netherlands (Fokker) with a later and less successful entry of the Ford 4 A-T Trimotor built by the Ford Motor Company of America. The planes were slow, noisy, and assaulted one's sense of comfort, but they represented a large step forward in luxury when compared with open-cabin predecessors.

The pioneer years of air transport continued through the decade as Austria, Finland, Hungary, Thailand, Russia, Czechoslovakia, Mexico, Italy, Bolivia, and Iran introduced passenger service. Most of the routes were operated only in the summer, in clear weather, and night service was gradually attempted. Air-cooled engines replaced water-cooled engines. Radio came into increasing use. Airport lighting was developed, and Germany, in particular, gave considerable attention to finding ways to navigate and land in inclement weather.

The growth of air transportation reached a remarkable size by 1926. The airlines operated by Germany, England, Sweden, Norway,

Denmark, Switzerland, and France were flying more than 10 million kilometers annually and carried more than 100,000 passengers. Although in many respects the U.S. aviation network was superior to Europe's, no American airline offered scheduled passenger service.

In the late 1920s, the first flights began in Peru, Yugoslavia, and Brazil. The year 1927 represented another historic milestone in commercial air travel. It was then that the world became air-minded as the "Lone Eagle," Charles Augustus Lindbergh, made his daring solo flight across the Atlantic. During that year, Pan American World Airways began operating scheduled international seaplane service while other modest seaplane operations were introduced in Brazil with Junkers. The *Graf Zeppelin* flew around the world; weather information was transmitted by teletype; welded steel fuselages came into use; and Jimmy Doolittle made the first successful instrument landing. There was a steady development of new standards of comfort and safety. For the next few years, enormous growth took place in air transportation. Junkers and Fokker began to supply trimotor aircraft the world over. The all-metal machines were becoming steadily more popular on all routes, all but replacing flying boats.

Although ground facilities were limited, harsh climatic conditions existed, and trimotors were low-performance, European countries were still eager to establish air links with their overseas territories, and this was to lead to the opening of trunk routes, which finally grew into the present global system uniting remote corners of the globe. Though some flying boats were still being used, by the end of the 1920s, Junkers and Fokker were offering all-metal construction, trimotor reliability, and the luxury of an enclosed cabin. Both manufacturers were supplying trimotor aircraft the world over, and impressive growth was achieved in revenue passenger kilometers (RPK). The all-metal air machines were setting aircraft standards of excellence and were becoming increasingly more popular.

Service started in Indonesia and Bulgaria in 1928. During the next year, airline flights were offered in Chile, Cuba, Union of South Africa, and Venezuela.

During the period from 1930 to 1936, the trimotors continued to dominate the world routes while flying boats were reduced to minor rivals. The reliable trimotors continued to capture sea routes until the flying boats dwindled to limited usage on long oceanic flights. In 1930, Gibraltar, Korea, and China could boast of first passenger flights.

Lowell Yerex formed an airline in Honduras in 1931, the same year that startups occurred in Rhodesia, Thailand, and Vietnam. In the middle of the Great Depression of 1932, Egypt introduced its first commercial air service. Even though 1933 was a Depression year, J.R.D. Tata, a wealthy Indian from Bombay, began operating a first-class air service in India. In that same year, airline flights were begun in Turkey. In 1934, Guatemala began a commercial air service.

EARLY AIR SERVICE IN THE UNITED STATES: AIRMAIL

While most of the world was building government-subsidized passenger air service after World War I, the United States elected to subsidize an airmail effort instead. Air service was costly, passenger service especially so, and it was not thought an economically sound endeavor in the early 1920s. And by 1924 all eight flying-boat airlines in the United States, including the first regular international service, had failed, thus demonstrating the unprofitable nature of the airline passenger business at that time.

In 1920, production of commercial aircraft began in the United States with the aim of making air service less expensive. Still, the United States was far behind European design and manufacture.

The U.S. Post Office Department (USPOD) airmail project had fared little better than the passenger services. The problems of airmail service were too complex to be met by unrelated small carriers, even though some of their strategies and attempts to solve problems were to provide the pattern for the present worldwide air transportation network. However, public interest in the development of airmail service was strong and demand for the service persisted. In 1924, the first transcontinental day and night regular airmail service was inaugurated.

The Kelly Act of 1925 was the result of legislative support for air transportation. This bill was called the Contract Mail Act, and it took the flying of mail away from the government and made mail contracts available to private operators. The act laid the foundation for the establishment of a viable air transportation industry in the United States by enabling air carriers to receive sufficient income from the cartage of mail to stimulate the passenger and air freight business. Proponents of the bill were aircraft manufacturers, fixed-base operators, and government officials anxious to promote national defense. Even railroad tycoons pressed for the acceptance of this bill. They were aware that it might assist the growth of private airmail carriers, but privately they believed that the air carriers had no chance of succeeding. The rapid expansion and continued dominance of commercial aviation in Europe was a principal incentive to those who wanted the United States to become the world leader in the industry.

Shortly after these airmail contracts were distributed to private operators, the Ford Motor Company introduced the first scheduled air freight service and the first domestic passenger service (between Los Angeles and San Diego) under the Kelly Act. In addition to these scheduled flights, Ford was the manufacturer of the Ford Trimotor aircraft, which served on these market segments. The design of this all-metal airliner was copied from the successful metal Junkers (of German manufacture) and Fokker (from the Netherlands) trimotors. The Ford Trimotor (Ford 4 A-T) was the first U.S. airplane that showed any hope of turning the airline industry into a passenger operation. The Ford 4 A-T was an inefficient, fat-winged, washboard-sided (corrugated) airliner that cruised at 85 knots. However, it was powered by three engines, strutless, and comparatively streamlined, with an all-metal construction that appealed to passengers. Despite its inefficiencies, the Ford Trimotor provided important safety advantages (three engines, all metal, and sturdy construction) over existing airliners (one or two engines, fabric and wood, weaker construction) flying commercial routes in the United States. America began to flex its muscles and challenged the technical dominance that the Europeans had held since 1910.

The Air Commerce Act of 1926 is often described as the birth of the United States airline system, as the government reentered commercial aviation. Though the army and Post Office Department had been instrumental in the development of the airways and, before the end of the year, had completed a night airways system from coast to coast, the act transferred to the Department of Commerce the authority to build, operate, and regulate the nation's air routes. In effect, the government was back on the scene as a regulator of those airlines created by the Kelly Act of 1925.

Between 1925 and 1926, the 290 airmail operators increased to 420, and nearly 300 new airplanes were added to the country's commercial fleet. Few were to survive. Only 16 operators held contracts for mail by 1926. (By 1931, just one of the original operators was still flying.) In their first year of private operation, the airlines carried 6,000 passengers, losing money on each one. (Today, airlines board that number of passengers every few minutes.) America had a $6 million industry with little attraction for private investment capital. By comparison, the government-subsidized airlines of Europe continued to expand with new equipment and added routes.

As U.S. airline systems developed, two faults with the Kelly Act emerged: (1) payment was made only for carrying mail; and (2) awards went to the lowest bidder who, in fact, was often unable to perform satisfactorily. The Kelly Act was amended in 1928 to solve these problems.

In 1929, the United States signed the Warsaw Convention Pact, establishing rules and limits of carrier liability in international air transportation with regard to passengers and property. This agreement tied the international aviation community together.

The Watres Act of 1930 marked another turning point in the history of airmail service and the development of air passenger transportation. This legislation provided for subsidizing of unprofitable, but essential, passenger services by the U.S. government, revised the basis of mail payment, and gave the Postmaster General enormous powers to revamp the nation's airmail routes. Progress continued; stewardesses (nurses) flew on aircraft for the first time. Two-way radio was installed in a ground-to-air system. Three transcontinental airmail routes were awarded.

The Airline Pilots Association (ALPA) was organized in 1931 and the first U.S. commercial four-engine flying boat (Sikorsky S-40) began airline service on international routes.

THE BOEING B-247 AND DOUGLAS DC-3: MODERN AIRLINERS

Trimotor aircraft maintained dominance until 1934 when the Boeing B-247, the first modern-type airliner, appeared. The B-247 was faster than most fighter planes had ever been and was capable of carrying 10 passengers in comparative luxury. The twin-engine airliner rendered obsolete the trimotors with their slow torture by noise. Further, the Curtiss Condor, the last American effort at a biplane transport, never got a chance as the sleek new airliners took to the skies. The Boeing B-247 had such refinements as a retractable landing gear, control surface trim tabs, variable pitch propellers, automatic pilots, de-icing equipment, and was expected to revolutionize air transportation.

The Boeing B-247 did not capture the success it had programmed, however. The pilots at United Air Lines rejected the proposed Pratt & Whitney Hornet engine in favor of the lower power Wasp. With the Hornets, the B-247 could carry 14 passengers, but with the Wasps, the passenger load was limited to 10. This caused a setback for Boeing and stimulated Douglas Aircraft to develop the DC-1, which was soon strengthened into the DC-2. The DC-3, the plane that is generally credited with turning around the industry from years of financial loss to eventual profit, was a derivative of the DC-2 that was requested by American Airlines in 1934. American wanted a larger aircraft that could accommodate "sleeper seats." The result was the Douglas Sleeper Transport (or Douglas DST), which could be configured to contain 16 sleeping berths or up to 24 seats in a "day version." It was the day version of the aircraft, designated the DC-3 and introduced into service in early 1936, that proved to be the longest-lived aircraft in history.

The DC-3 could carry 24 passengers at a speed of 273 km/h over stage lengths of 800 kilometers. It not only increased the speed and comfort of travel, thereby winning passengers who had not been willing to brave an airliner before, but it also performed reliably and profitably. It can be said that it was the Douglas DC-3 that introduced the concept of operating costs to the airlines. The DC-3 was recognized as the first airplane to instill a feeling of confidence in air travel, as measured by the fact that its safety record made feasible the first air travel insurance for passengers.

More than 13,000 DC-3s were ultimately produced. Also seeing widespread military use, these aircraft were designated the C-47 by the U.S. Army Air Corps, and the R4D by the U.S. Navy. In 1990, some 56 years after their introduction, more than 500 DC-3s were still flying worldwide.

THE DOMESTIC AIR TRANSPORTATION SYSTEM

While the United States had begun to assert itself and challenge the Europeans in the area of aircraft production, the air service in America was plagued with problems. Airmail was still the center of U.S. attention in the air, but because of some irregularities perceived by government investigation in 1934, President Franklin Delano Roosevelt cancelled all of the airmail contracts and nullified the Kelly Act. He ordered the U.S. Army Air Corps to transport all airmail. This proved a virtual disaster, for the military was ill-equipped and without training. The resulting service was poor and the pilot fatality rate unacceptably high. Within a few months, the Air Mail Act of 1934 was enacted and authorized the government to award one-year contracts to private airmail carriers, subject to review later. The USPOD was given broad authority to award contracts and determine route allocations. The Interstate Commerce Commission (ICC) fixed the rates; the Bureau of Commerce was responsible for licensing and the regulation of airways.

At this time, aircraft manufacturers were required to divest themselves of airline ownership. Air carriers were given greater latitude in self-determination to pursue economic goals, and the Air Mail Act also created the Federal Aviation Commission to make policy recommendations. During the next few years, airport construction increased, ground facilities were improved, and airways control and equipment were upgraded substantially. As the system became better developed, passenger traffic increased beyond the most optimistic forecasts. Schedule performance reached new levels of reliability. Although passenger service was increasing in the United States, it was not until 1940 that passenger revenues exceeded the income from airmail payments.

The Great Depression had a striking negative impact on the air transport industry. The concept of open competition supported by subsidy led to irrational competitive actions. Bankruptcies occurred and the system was in turmoil. It culminated in the Civil Aeronautics Act of 1938. Under this act, the responsibility for the development and operation of the airways was assigned to the Administrator of the Civil Aeronautics Authority, later called the Federal Aviation Administration. This legislation also established the Civil Aeronautics Board (CAB) and the regulatory structure of the industry. The eventual result was a healthy airline system.

WORLD WAR II

Prior to World War II, several more countries around the globe had introduced air service. In 1936 it was Ireland, the Philippines, and Uruguay. In 1937 Iceland, Mozambique, and Saudi Arabia followed. Both Angola and Iran had services started in 1938 and the British West Indies inaugurated commercial flights in 1939.

In just two decades air transportation had made enormous progress. Airport construction, airways development, and facilities were improved substantially. The first airway traffic control center was established at Newark, N.J., to monitor and control all air traffic.

As was typical, however, it was the military applications and warfare that most strongly impelled advancement in aviation. The British installed and tested the first transponders in military aircraft, and a totally automatic landing system was also successfully tested. Passenger traffic increased beyond the most optimistic forecasts. In Europe just prior to the war, Adolf Hitler engaged his designers in aircraft research, which resulted in remarkable improvements in performance and reliability. Rocket power and jet engines were successfully produced.

World War II began in September 1939. Aircraft technology made another quantum jump in design, performance, and reliability. Whereas the airplane was prominent in World War I, it decided the course of World War II. Because the commercial air carriers contributed a crucial share to the war effort, air transport research in logistics, economy, and performance was stressed. Air cargo increased, passenger traffic multiplied, and the global nature of the fighting led to the wide use of air transports in international travel. As a result, Allied nations became not only air minded, but internationally air minded.

Aviation in all its forms was nothing if not a huge weapon to be mass produced. By Allied agreement, the U.S. production effort emphasized transport aircraft, developing expertise that was transferable to the postwar period.

Because transcontinental and transoceanic capabilities were essential to the war effort, emphasis on development and design of longer-range aircraft continued. The standard for long-range, high-performance air transport operations centered on the Douglas DC-4. This four-engined aircraft carried a crew of six, more than 40 passengers, at 320 km/h, and with a range of 2,400 kilometers.

In addition to the DC-4, Lockheed developed its Constellation aircraft during the war. Also a long-range, high-performance aircraft, the Constellation helped prepare the world for the coming Jet Age.

THE POSTWAR YEARS

Air transport development, which might otherwise have taken a generation, was telescoped into a wartime period of a few intense years. The enormous wartime production of transport aircraft flooded world markets after hostilities ended, making air travel across continents and oceans as simple as by rail or sea. During the war Boeing had developed the first pressurized airliner that would permit flying over the weather. Christened the *Stratoliner,* the four-engine craft did not find wide use and not many were built.

War surplus DC-3s and DC-4s continued to be sold to civilian airlines all over the world to rebuild war-torn networks of passenger service, start new airlines, or launch air cargo ventures. In the first postwar year, worldwide airline route kilometers increased 23%, passengers carried ballooned 82%, freight rose 61%, and mail jumped 52%. Most of this growth was accomplished with war surplus equipment.

It is notable that the war proved to be such a boon to air cargo development. Military support was so pivotal to the war effort that new tactics in logistical cartage were being constantly developed. The complicated operations of war also enhanced the growth of radio communications. Radar became a high-priority project, later to develop into the cornerstone of air traffic control. Military air traffic in high-density environments became valid models for sophisticated air traffic control techniques.

During the early postwar years, the international airlines experienced tremendous expansion without achieving accompanying profits, however. The surplus aircraft, although purchased at favorable prices, still had incredibly high costs of operation.

A new breed of pressurized airliners began to supplant the older, war-surplus aircraft: the Douglas DC-6 (1946), Lockheed L-49 Constellation (1945), Boeing Stratocruiser, and Vickers Viscount (1948) dominated the international airways.

During the war, it seemed that Great Britain concentrated on the production of fighters and short-range bombers whereas America took on the task of heavy, long-range, four-engine bombers and high-performance logistical support aircraft. Ironically, Great Britain's postwar position was that, although the nation led the world in jet propulsion, it had no experience in large transport applications. At the termination of hostilities, therefore, Britain had few aircraft convertible to commercial flying even though British technical progress had been conspicuous. One British aircraft did reach widespread usage, however: the four-engine Vickers Viscount.

While several equipment manufacturers were vying for control of the skies, several new countries inaugurated air service for the first time. In 1944, Guatemala formed its first airline. In 1945, time lost during the war was made up as service was introduced in Algeria, Ethiopia, Morocco, Madagascar, Lebanon, Hong Kong, Iraq, Pakistan, Romania, Sri Lanka, and the Sudan.

Trained personnel, excess spare parts, and operating experience made the transition relatively easy. Air travel across oceans and continents became routine. Unfortunately, however, during the war the central focus on the development of large aircraft was the load capacity and not economy of operation. It was not unexpected that the tremendous surge in postwar civilian airline revenues did not result in a profitable industry, because of the high cost of operation. A basic lesson in airline economics was again being learned, and that was the importance of direct operating costs. Most of the aircraft went from mufti to civilian garb cheaply and, therefore, with excellent revenue-generating capabilities, but with excessive operating costs that swallowed up profits; thus, the government had to come to the rescue with subsidies to support the losses.

In 1947, airlines began limited operations in Cyprus, Korea, Syrian Arab, and Ecuador. During the next year, more countries took to the air with national flag carriers; these included Burma, Israel, Tunisia, as well as two international nonscheduled operators in the United States, namely TIA (Trans International Airlines) and World Airways.

THE JET AGE

After several ill-fated attempts to produce propeller aircraft to compete with those of the Americans, the British began tapping their expertise in jet propulsion gained during World War II. Britain was the first country to produce a pure jet airliner in the de Havilland Comet I.

The year, 1952, was an historic one. Her Majesty's Government proudly launched the de Havilland DH-106 Comet I, the aircraft that was first to herald the Jet Age. Scarcely two years after its introduction, tragedy struck the Comet. Constant pressurizing and depressurizing of its metal cabin produced minute cracks in the hull due to metal fatigue, a concept that was poorly understood by aircraft manufacturers in the early 1950s. In 1954, two BOAC Comets exploded over the Mediterranean with all on board lost. Within three weeks a third Comet disintegrated in flight, and the aircraft was withdrawn from service. The graceful flagship of Britain's jet fleet was grounded, never again to carry passengers in its original form. By the time the vastly modified Comet IV entered service to take the place of its predecessor four and a half years later, the Soviets and the Americans already had jet aircraft flying.

The distinction of being the second country to offer jet service went to the Soviet Union, which ushered in the TU-104 jetliner in 1956. The TU-104 was also the world's first swept-wing jetliner, adapted from a Russian bomber. It was not until 1958 that the first American jetliner, the Boeing 707, finally took to the skies across the North Atlantic. An interesting sidelight is that engineers at Douglas Aircraft Company designed the competitive DC-8 passenger jet beginning two years after Boeing, and the prototype DC-8 was completed first. This is a true testament to the engineering and technical skills at Douglas.

Though the DC-8 proved to be an excellent jet, it never quite attained success on par with the Boeing 707. Both American jetliners carried more than 170 passengers, at speeds in excess of 800 km/h, nonstop over distances of 6,400 kilometers, and at altitudes of about 10,000 to 13,000 meters. The DC-8 operating costs were, however, about 5% to 10% greater than those of the Boeing 707. The Jet Age was upon us, as turboprops were rendered obsolete and airspeeds increased by 50% at a stroke.

Airline and airway legislation continued throughout the postwar years in the United States. In 1946, the Federal Airport Act stimulated the construction and development of airports, and the Burmuda Agreement, which served as a model for subsequent negotiations, delineated rights for the operation of each country's commercial airlines over and into the territory of the other. In 1948, the first trunk carrier went off subsidy, and some 27 airlines were authorized to operate feeder routes. By 1951, American, Eastern, TWA, and United, considered the "big four," were all off federal subsidy. Congress passed the Federal Aviation Act of 1958, and in the same year created a new Federal Aviation Agency (FAA) as one independent and comprehensive governmental agency to control all aviation matters, both civil and military.

Instrument-landing systems were standard by the late 1950s. Air navigation was improved with the invention of the transistor. Radar coverage of the skies, or positive traffic control as it is referred to by the airlines, was implemented.

At the same time that the jets were impressing most of the aviation world, during the second half of the 1950s, airline service began in Afghanistan, Cambodia, Ivory Coast, Yemen, Libya, and Nepal. Australia's Qantas began the world's first round-the-world scheduled service with the Lockheed Super Constellations.

While the transition to jet fleets was taking place, the turbine-powered Vickers Viscount continued to be in demand until 1959 when the French Caravelle (with engines in the rear) was available as a short-range jetliner. Before the end of the decade, almost every major country provided air service with its own national flag carrier. There had been a threefold increase in the number of aircraft in service around the world during the 1950s. The average trip length of international carriers doubled in distance between market pairs, and the industry soared to new revenue heights. The 11 airlines criss-crossing the North Atlantic inaugurated Tourist (economy) service for the first time. Commercial flights over the polar region between Scandinavian countries and North America were pioneered by SAS, still using propeller-driven Douglas DC-6 aircraft.

Although in the 1960s, propeller and turboprop aircraft continued to serve traffic demands, jetliners began to dominate the bulk of airline flying routes. The aft-engine vogue brought forth designs of two, three, and four engines mounted in the rear of the fuselage. The principal twin-engine craft were the French Caravelle, British BAC-111, and, especially the Douglas DC-9, the most widely used twinjet of its time. The trijet formula was exemplified in the British Trident, Russian TU-154, and the Boeing 727, which was, until recently, the most successful aircraft ever built. The American aircraft captured the world markets. As a matter of fact, the Boeing 727 proved to be the largest selling jet airliner in history until a recent purchase of Boeing's newer twinjet, the 737, broke the long-standing record.

During the 1960s, Malaysia and Singapore initiated air services. Also, TWA announced that it was the first major airline in the world to have converted to an all-jet fleet. The supremacy of narrow-bodied jets over the air routes of the world was complete, and only comparatively few propeller aircraft continued in service over 1,000-kilometer distances.

Aircraft designers began work on a new type of aircraft in the 1960s. The supersonic airliner was researched around the world, but judged infeasible by American designers and the program was abandoned. An Anglo-French consortium, Aerospeciale, however, pursued the new technology after receiving government support in 1963.

The Transportation Route Investigation in 1966 permitted several U.S. carriers to enter the Pacific Basin market. The following year, a transatlantic case was instituted to determine whether new or improved service was necessary to relieve congestion at the New York airports. Mergers began to take place in 1968 when five airlines merged in Alaska.

THE WIDE BODIES

If the 1960s represented the decade of the narrow-bodied (single aisle) jets, the 1970s represented the high time for the wide-bodied jets (two aisles). To cope with the increasing demands of mass travel, the "jumbo jet" Boeing 747 was developed and introduced into airline service on January 22, 1970, on a Pan American flight from New York to London. The enormous aircraft featured two parallel aisles up and down the cabin and could accommodate up to 500 passengers in an all-economy configuration with 10-abreast seating.

The new Boeing 747 was followed one year later, in 1971, with the McDonnell Douglas DC-10 and Lockheed L-1011 wide-bodied trijets. The European consortium's A-300 Airbus was introduced in 1973 and, after a slow sales start, found relatively notable success and use as the first wide-bodied twinjet.

Initially, the wide-bodied jets provided excess capacity for the amount of traffic available. After this introductory phase, however, these aircraft proved to be particularly efficient and economical. The airlines committed more than their entire net worth to buy new jets, which outclassed the older propeller aircraft in every way: jets could fly twice as fast, at altitudes twice as high, and could carry twice as many passengers. Economies of scale were also notable, and the cost per passenger seat kilometer (PSK) was dramatically reduced. For the first time since the introduction of the Douglas DC-3, the airlines could expect to utilize a new aircraft through its entire depreciation life (about 12 years). With the cost of production of new jets rising at an annual rate estimated at 8.8%, the result was that each aircraft maintained a residual value in excess of its purchase price.

The fuel crisis of 1974 sent the industry into a tailspin. No longer was jet fuel inexpensive. Worldwide, carriers found their fuel costs multiplied 10 times over.

The inaugural flight of the new Aerospeciale Concorde Supersonic Transport (SST) (from London to Bahrain) took place on January 21, 1976. The 2,300 km/h, 125-passenger-capacity Concorde attested to the engineering genius of Britain and France, the two countries that jointly built the SST, but the aircraft proved to be a technical masterpiece and an economic failure. Though 15 aircraft were produced, only 10 are still flying. Both Air France and British Airways (which operate the Concorde) are limited to much less capacity because of fuel requirements. Thus, load factors described as high are somewhat misleading as the percentage refers to the lesser capacity (80 to 100 seats) and not to the 125 available seats for which the SST was intended. Additionally, various governments placed restrictions on flights over their countries, demanding subsonic speeds (because of sonic booms). With the somewhat limited range and imposition of subsonic speeds over land, a flight from London to Sydney in the Concorde was about four hours faster than in a Boeing 747, hardly a supersonic advantage!

With the exception of the Concorde, completely new aircraft designs were slow in appearing. Without the direct participation of governments, the cost of developing new aircraft is almost beyond the resources of even the largest aircraft manufacturers. The result is the wide use of aircraft derivatives. If a jetliner proved to be successful, it was modified to perform as many airline requirements as were feasible. The Boeing B-747SP (Special Performance) was the ultimate example. Its maiden flight was on July 4, 1975, four and a half years after the first 747 took to the air. The 747SP was 15 meters shorter than the conventional 747, with a range thus far unsurpassed. Carrying up to 400 passengers in the maximum, high-density configuration, the 747SP could operate nonstop from New York to Tokyo, from San Francisco to Hong Kong, or from San Francisco to Auckland.

AIRLINE DEREGULATION IN THE UNITED STATES

By the mid-1970s, however, regulations and political pressures had become so cumbersome that rational efforts to reduce constraints upon the air carriers proved to be futile. Finally, in 1978, the Congress deregulated the air transport industry. Suddenly, airlines were able to increase their route mileage, reduce fares, and operate in a more competitive fashion. Changes were massive, chaotic, and rapid. Fares were reduced sharply, schedules increased, and the industry strove to become a mass transit air system, but without a master plan. In 1932, U.S. airlines operated 700 flights daily. Immediately after deregulation they operated that many each hour, in aircraft carrying up to 40 times more passengers, at six times the speed, and flying five times as high. The United States had entered the era of mass air transport.

From the international perspective, deregulation in the United States in 1978 had a deleterious effect on the world's routes for travel. Passengers became far more cost-oriented than they had been in the past, and the airlines were forced to be more cost-conscious as well. The result was that, at the end of the decade, traffic fell to new lows and international air carriers found themselves engaged in no-win marketing battles for limited passengers in an over-capacity environment and declining demand.

This brought a metamorphic change in the manner in which air transportation business was conducted. Operating costs reached new highs. Competition was heavy with excess capacity and reduced yields. Planes were maturing, and it was estimated that, over the next 10 years, there would be equipment requirements in an amount exceeding $100 billion for an industry without earnings and a disillusioned investment community. New aircraft were economical to operate but extremely expensive to buy. The margin between debt service and economical efficiency was narrow and funds difficult to obtain.

New-generation, high-efficiency aircraft could provide a competitive edge to the airlines that possessed them. Knowledge of this prompted Boeing to consider production of a whole new family of aircraft in early 1978. Though most of the aircraft entering the air transportation market are derivatives, the new Boeing 757 and 767 twinjets, which both began flying in the early 1980s, represent a completely new generation of aircraft. These planes have been proven to be 30% to 40% more efficient with a new wing design and substantial improvement in engine performance. It has been estimated that Boeing committed the equivalent of its entire corporate resources to develop this new family of aircraft.

Within the next decade, it is believed that the domestic airline industry will have to replace its entire fleet. This will require a capital investment requirement far in excess of the airline industry's ability to fund using traditional means. It is likely that the federal government will have to underwrite a portion of this debt.

Over the years, there has been an inexorable movement of the airline industry to a mass-transit operation. It is likely that the industry will hold relatively stable in the near term. In order for quantum leaps to occur, it will require technological changes unsupportable by the industry at this time.

The air transport network has evolved from a single aircraft to an industry. In 1979, more than 80% of all intercity first-class mail went by air; U.S. airlines served more than 600 airports and now carry approximately 10 billion kilograms of freight annually. About 50 airports in the United States accommodate more flights daily than London's Heathrow Airport, Europe's busiest.

It is improbable that air transportation will ever return to the simple form it once had. The industry is undergoing turbulent times. Many experts predict that it will characteristically experience increased maintenance problems, a period of some questionable safety incidents, fuel-price increases—which will seriously impinge upon the profit of the industry—marginal carrier operations, and many mergers. Eventually, however, a functional air mass-transit system will prevail.

Chapter 2

The Domestic Air Transport Industry

Courtesy of Fokker Aircraft; Fokker 100

STRUCTURE

Deregulation of the airline industry opened the gates; more than a hundred new carriers emerged and many operating carriers restructured their route systems dramatically.

Because the historical industry groups had lost their significance, the CAB defined new categories for the U.S. carriers: Majors, Nationals, Large Regionals, and Medium Regionals. While the old, regulated groupings were based on historical entry into the market and size of their system, the new deregulated categories are determined by annual revenues. (The board considered many methods to classify the airlines for statistical purposes and concluded that this was the most objective criterion, though not a perfect one.)

Each of the four categories listed above is broken down into smaller classifications when appropriate. These are: Combination Service (passenger and cargo), All-cargo, and Charter. Currently, some examples of the carriers assigned to each group can be found in Figure 2.1.

MAJORS

(Carriers with annual revenues over $1 billion)

American	Northwest	United
Continental/Eastern	Pan American	USAir/Piedmont
Delta	TWA	

NATIONALS

(Carriers with annual revenues $100 million to $1 billion)

Air Midwest	Air Wisconsin	Alaska
Aloha	America West	Braniff
Hawaiian	Midway	Southwest

REGIONALS/COMMUTERS

(Carriers with annual revenues under $100 million)

Air Pennsylvania	ASA	Aspen
CCAir	Christman	Comair
Direct	Florida Express	Henson Horizon

Metro	Midway Commuter	
Pan American Express	Reeve Aleutian	Simmons
SkyWest	WestAir/NPA	
(Just to name a few)		

Figure 2.1   Major/National/Commuter Airlines As of June 1988

COMPETITIVE VARIABLES

During the years of air transport regulation, whether or not Competitive Variables could produce meaningful results had been an open question. The constraints placed on competition by government controls were severely limiting. An airline could not change its route structure, pricing, or its basic operating techniques without CAB or FAA approval. Additionally, proposed route and pricing changes had to be filed with the CAB. Detailed operating and financial statements had to be submitted. As regulated carriers, all information reported to the CAB became public domain; thus, competitors had free access to everything.

With deregulation in 1978, Competitive Variables assumed roles of major importance. Although the typical life span of any innovation is often brief, innovations can increase market share, raise load factors, and improve an airline's economic performance.

There are three major Competitive Variables along which an air carrier can differentiate itself from its competition. These are: schedule, routes, and pricing. There are six of a minor scale: frequency, equipment, service, convenience, loyalty, and perception. Each will be examined in detail.

Schedule

Airline planners strive to maintain a delicate optimal balance between operations and marketing. In this context, *operations* strives to maximize utilization of aircraft related to hours per day, minimum down time, and low operating costs. *Marketing* revolves around peak times, load factors, and objectives. The interaction can be dramatic. An airline that originates only a few flights on market pairs in order to increase load factors might quickly lose many passengers to competitors providing capacity and frequency. On the other hand, empty seats cannot be inventoried and can plunge an airline into a negative cash flow. As utilization of equipment increases, however, delays are more apt to occur.

Schedules provided between market pairs might be related to

schedules on other segments. For example, to maintain a minimum schedule of two flights each day along a high-density route, an airline might have to tolerate low-load factors and a small market share. Or, to accommodate connecting passengers on direct routes, schedules between city pairs might have to be increased to three flights, thus reducing the load factor per flight. There may be a requirement to position an aircraft for the next day or schedule maintenance at a facility, requiring an off-peak flight with a low-load factor.

Schedules can be influenced by peaks in demand during certain times of the day, week, or year. In particular, early morning, noon, and early evening flights are more in demand than mid-morning, early afternoon, or late-night flights. Holidays have a tremendous impact on schedule needs and are often in one direction with no support for return flights.

The symbiosis of matching these high-demand periods with low-demand times is a complex challenge for schedule planners. When unforeseen circumstances occur such as flight delays, cancellations due to component failures, or below minimal weather, the problems are compounded. To achieve high hours of utilization, aircraft are scheduled for short transits. A delay can disrupt the entire day for that aircraft, have a system impact, not to mention the workload of both ground and flight personnel.

There is no question that the most successful air carriers have selected scheduling as the cornerstone of their planning strategy. Scheduling alone can establish or destroy customer preference. While most of the scheduling can be done mechanically, the critical choices are more often made based on the informed judgment of management. It has been established that disruptive circumstances occur in about 10% of all flights.

Thus, scheduling represents the highest-order integration of most of the functions of the airline: selling, advertising, market research, pricing, service, origin and destination (O&D), traffic peaking, and utilization.

Routes

An air carrier's route network is a vital consideration. With the dynamics of deregulation, the route system has become a vulnerable Competitive Variable. The ease of ingress and egress has become a volatile element in route structures of airlines. Start-up costs are heavy considerations for airlines in the determination of entering new routes and, in addition, if a route proves viable, it is likely that competition will enter the market.

In the matter of market pairs, a flight segment may be Nonstop (no landing), Direct (landing between O&D) or Interchange (change of airline or flight required). Route networks present new problems. There are three general types that are often commingled. One is the *Hub-and-Spoke* system with a major hub in the center and routes radiating from that hub (see Figure 2.2).

Second is the *In-line* route structure with service being provided between two points in a variety of subroutes (Figure 2.3).

Third is the *Grid* network, which follows the pattern the name describes (Figure 2.4).

Another strategy has developed in route determinations revolving around satellite airports instead of major terminals. Several airlines have established successful city pair segments by using peripheral airports, with attendant benefits. Figure 2.5 highlights some of the advantages and risks inherent in each route strategy.

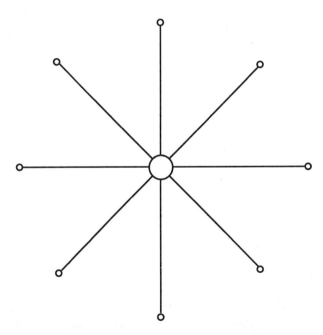

Figure 2.2 Typical Hub-and-Spoke Route System

Figure 2.3 Typical In-Line Route System

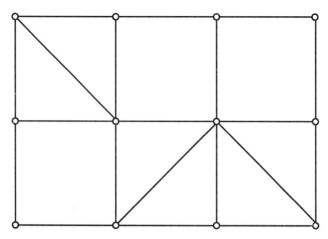

Figure 2.4   Typical Grid Network

Figure 2.5   Advantages and Disadvantages of Each Route Strategy

	Hub and Spoke	In-line	Grid
Marketing/ Operation	Centralized	Decentralized	Quasi- centralized
Peaking	Sensitive	Quasi- controllable	Controllable
Directional imbalance	Severe	Potentially severe	Controllable
Bad weather	Severe at hub	Severe	Local only
Utilization	Low	High	Controllable
Load factor	Low (usually)	Low	Controllable
Station density	High	Low	High
One-type service	Quasi- controllable	Controllable	Quasi- controllable

Pricing

Of all of the Competitive Variables, pricing is perhaps the most complicated and transitory. It is the art of translating into quantitative terms the value of the journey to passengers at a point in time, relative to competition.

Pricing involves a variety of classes, discounts, group charters, and time of day. These change on a daily basis, and thus it is extremely difficult to maintain a price differential between airlines. That is not to say that pricing is inelastic. Yet there is an almost continual downward pressure on fares as one airline attempts to gain some temporary competitive advantage.

In general, elasticity increases with flight segment length. Pricing is more apt to be inelastic on shorter routes owing to aircraft being of smaller size, higher costs per mile because of frequent landings, and lower load factors.

Pricing is the most intriguing of the variables. Management personalities are manifested strongly in pricing behavior.

A rational pricing/load factor strategy is imprecise, yet it is mandatory in estimating price/load factor relationships in making pricing decisions. In the absence of quantitative techniques for determining the level of demand for a given price, it is helpful to understand the relationship. The most illustrative is the construction of a demand curve, which is a subjective evaluation of such a relationship. Its key contribution is as a concept in thinking about setting prices and also in analyzing, on a continuing basis, market reaction to price changes (see Figure 2.6).

Frequency

Time of departure is a vital consideration for passengers. An airline that maintains frequency, rather than capacity, is more apt to capture a large market share. However, to maintain frequency requires smaller aircraft with less capacity and a higher Direct Operating Cost (DOC) per seat kilometer. Under certain conditions, planners place an aircraft into a market pair to increase utilization although it provides

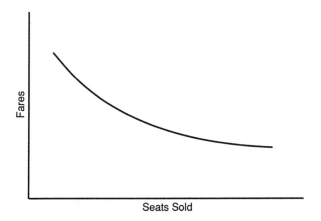

Figure 2.6 Seats Sold vs. Air Fares

excess capacity. Planners must make the determination of the trade-off of utilization versus excess capacity. Added to that dilemma is the growth factor. That is to say that an excess capacity aircraft will be instantaneously responsive to growth and new peak demands. Frequency will permit straddling a peak period. A carrier operating a Boeing 747 may serve the apogee of the peak demand time.

Low frequencies are less elastic in both the cancellation or addition of flights, as well as schedule flexibility.

Frequency is a powerful tool for disrupting passenger loyalty since it is common knowledge that passengers with strong airline loyalties often abandon them to take the first flight out, using subsequent flights as backups. In the present days of "frequent flyer" benefits, the ability to disrupt loyalty is an important Competitive Variable indeed.

Equipment

The type of aircraft that an air carrier operates is a variable upon which many industry executives place the greatest emphasis. Passengers are believed to prefer the newest, fastest, and most comfortable aircraft available if given a choice.

Many airlines believe that they must match or surpass their competitor's equipment or lose market share. Additionally, there is a pressure on the upside to provide excess capacity because of overbooking, growth perceptions, and traffic loss. Delta and Singapore Airlines are classic examples of the efficacy of operating the latest equipment.

Service

When the above Competitive Variables are comparable, customer service takes on a role of great significance. The air traveler buys not only a seat on an aircraft but all of the services provided. The passenger may place a high value on the name of the air carrier, the various support systems such as hotels, tour packages, rent-a-car facilities, movies on board, and other options. From the moment a passenger makes an initial contact, there are numerous opportunities for product differentiation.

Ticketing procedures, information check-in facilities, baggage handling, boarding, in-flight meals, flight attendants in action, off-loading process, and baggage retrieval are a few of the service functions that can delineate carriers.

Convenience

One of the reasons off-center airports have appeal is the convenience in parking, shorter lines at the airport, less distances to walk, and, more often, lower parking fees.

The difference between San Francisco Airport and Oakland, O'Hare and Midway, JFK and Newark, Heathrow and Gatwick, and other pairs is enough to influence the carrier selection, in spite of the probability of less flight selections.

Loyalty

Through the medium of advertising campaigns or passenger experiences, loyalties develop. One air carrier makes its primary thrust in the area of safety, which it may view as a primary concern of passengers. Another caters to the business traveler and generates a cadre of followers. Loyalties develop for an array of reasons. Given common factors for a flight decision, passengers tend initially to select the air carrier to whom they are loyal. Airline "frequent flyer" programs have done a great deal to establish airline loyalty.

Perception

Perception is the sum of the variables as perceived by the passenger. The passenger identifies an air carrier as having the most desirable schedules, most frequencies, lowest prices, best service, newest equipment, and is loyal to that carrier. It may be factual or imagined, but it is that passenger's perception which is a strong motivating force when buying a ticket.

There is an absence of replicated market profiles between pairs and the use of variables. To define clearly the optimal variable or attach a mathematical value to a variable demands an unwieldy specificity not available.

Thus, a manager must incorporate simplifying assumptions and general parameters to maintain real-world probability with the recognition that Competitive Variables are transient in nature and are most effective for short periods of time over limited geographical areas.

A multitude of varied socioeconomic and geopolitical factors, together with influences indigenous to the industry itself, combine to keep traffic flow cyclical, difficult to estimate and forecast, and market equilibrium in constant motion. Competitive Variables tend to change values and priorities in a microcosm of shifting traffic influences, either stimulants or depressants:

a. Traffic Stimulants

Economic prosperity	New routes
Seasonal demands	Schedule changes
Discretionary travel shifts	New aircraft
New destinations	Pricing tactics
Travel packages	Advertising

b. Traffic depressants

Business recession	Airport congestion
Additional competition	Reduced frequency
Market pair reduction	Increased fares
Political constraints	Military conflicts
Bad weather	Accidents
Travel shifts	Schedule changes

Summary

Competitive Variables offer four major methods (schedule, routes, pricing, and frequency) for product differentiation and five minor ones (equipment, service, convenience, loyalty, and perception). The optimal use is essential for the success of any airline, for every variation has a cost impact. The critical task facing management, therefore, is to determine accurately the cost/benefits associated with each Competitive Variable decision.

A tremendous amount of uncertainty is associated with the application of Competitive Variables because of the potential response of competitors. It is also true that value judgment plays a major role in the selection of a Competitive Variable as a product differentiation decision. It is most important to be aware that Competitive Variables are pivotal tools that can capture market share and profitability when employed correctly.

OPERATIONAL CHARACTERISTICS: AIRPORTS

In the United States, most public airports are owned and operated by local governments. In effect, they represent a subsidy to the users and are regarded as a form of local business development or advertisement.

The fee each airline pays for using airport facilities is usually individually negotiated with the local airport administration. At an increasing number of airports, income from charges to airlines is supplemented by fees from observation decks, view telescopes, concessions to service businesses located at the airport, and ground leases for air-transport-related business offices. Nevertheless, most airports are not self-supporting; government subsidies are still a necessity.

Terminal facilities at airports are crucial to the speed and convenience with which passengers can be served, and airlines normally lease sales and administrative space at the terminal and in other suitable locations in the surrounding communities. However, land, buildings, and ground support equipment ordinarily represent about one-fifth of an airline investment in physical assets; this is relatively small compared to flight equipment costs.

Most experts agree that airports must be continually upgraded to meet the demands of traffic. The airports' runways determine, by their length, number, and orientation, the potential volume of traffic and types of aircraft that can utilize the facility. When commercial jets first entered the market, only a few U.S. airports had runways long enough to permit the operation of these aircraft. Thus, an extensive program of airport expansion was initiated in nearly every major U.S. city and in the world. Furthermore, only rarely has an airport been built in the United States or anywhere else that, upon completion, met the requirements of the current traffic.

The Federal Aviation Administration (FAA) provides and operates facilities for air navigation and instrument landing equipment at most major airports. Control by the FAA assures a degree of uniformity in practice that would be difficult to achieve if each community were to establish its own facilities and procedures. Traffic control has become more sophisticated. Dual runways are often used for takeoff and landing. Radar is used in every stage of flight for surveillance and control, even during taxiing. With improved instrument landing systems both on the ground and in the aircraft, airlines are on the threshold of all-weather operation as indeed exists at some airports.

Aircraft

In 1988, aircraft and flight equipment accounted for about 80% of the average carrier's fixed assets. Spares inventory (and part for engines) represents about 20% of this investment, with the remainder in aircraft. The capabilities of an airliner dictates its operating environment, capacity, speed, and utilization.

Matching these capabilities to market demand is a critical strategic decision. The opting by one carrier to purchase newer, faster, or more comfortable aircraft places pressure upon its competitors over the same routes to follow the same fleet acquisition plan.

For example, the jumbo jet, wide-bodied Boeing 747 was initially acquired in 1969, and was uneconomical because it offered excess capacity. Therefore, though the cost per available seat kilometer (ASK) was lower than competitive airliners, because of its larger size and scale economy, the market was comparatively inelastic, and load factor dropped.

However, the wide-body jet threatened to capture a larger share of the market, and many airlines were compelled to buy B-747s to maintain their competitive positions. Most airlines would have preferred to continue operating the smaller-capacity B-707 and DC-8 aircraft until the market demand developed to support the B-747 capacity, which had room for more than 400 passengers.

Nevertheless, the jumbo jet was faster, offered an upstairs lounge, was less confining (owing to the wide body and two aisles), and gained immediate passenger preference. Only after several years did traffic demand meet the jumbo jet's capacity. At the time of introduction in 1969, the B-747 was the most cost-efficient aircraft in the air.

Derivatives of aircraft are subsequent improvements or modifications of the same model. Typically, an airliner model is produced and derivatives are found to be stretched models, more powerful engines, or simply refinements as a result of research and development (R&D) experience.

Some fleet-planning experts have suggested that the B-747SP (Special Performance) derivative introduced in 1975 might have served better as the basic model for the 747 model as it was 48 feet shorter, a rare derivative option. It had the capacity of the trijets, the DC-10 and the L-1011, in the range of about 260 passengers. The sequence would then have been for the B-747-100 to become the derivative of the 747SP and the fuselage extended to increase capacity as demand rose.

Several airframe manufacturers currently supply the U.S. commercial air transport industry. Through the years the four most important manufacturers have been

1. Boeing, which manufacturers the 707, 727, 737, 747, 757, and 767 jetliners
2. McDonnell Douglas, which manufactures the DC-8, DC-9, DC-10, MD-80, and MD-11 jetliners
3. Lockheed, which manufactured the L-1011
4. Airbus, which manufactures the A-300, A-310, and A-320 jetliners.

Three prime engine manufacturers supply jet engines for U.S. airlines: Pratt and Whitney (U.S.), General Electric (U.S.), and Rolls-Royce (Great Britain).

Whenever a manufacturer designs a new aircraft model, it is with close cooperation of the airlines. During wars, governments funded R&D, and airliners emerged as military derivatives applied to the air transport industry.

Ever since World War II, manufacturers have had to fund R&D for new-generation aircraft and have been willing to do so only with confirmed orders. The lead time for production varies from two to five years, and the manufacturer typically requires at least 200 orders to recover R&D costs.

Therefore, aircraft manufacturers are forced to seek the broadest sales base before committing to build an aircraft. Such a game plan is a barrier to entry and reduces the chances for the competition to suddenly introduce a newer, faster, or larger aircraft on its routes. Granted, the first airline to order and receive delivery of new aircraft does gain a competitive advantage until the competition obtains its new aircraft.

Ground Support

Fleet planners must determine what ground support will be required at various stations to provide adequate service. Ground-support equipment includes miscellaneous vehicles such as baggage carts, mobile ramps, commissary vehicles, and fueling trucks. This function is far more complex than appears, for it might be more economical to rent equipment for an airline that has only one flight a day into a station. Certain ground-support equipment may be flexible and can be used for an entire fleet of twinjets, trijets, and four-engine jets.

Some ground-support equipment may have growth potential that would allow for its continued use in new aircraft to be delivered. This cross-utilization, rent, lease, or buy decision is especially important because costs for ground-support equipment can vary between 10% and 25% of the cost of flight equipment (aircraft).

Rising capital costs of sophisticated ground-support equipment have stimulated joint service usage between airlines: One airline provides another carrier with its equipment for low-frequency service at incremental costs. Separate ground-service firms have developed, which are capable of supporting airlines with all required ground handling at reasonable charges. The use of such ground-service companies can be particularly attractive to airlines, for they do not require capital expenditure and allow flexibility of varying schedules and capacity of service with only short-term contractual constraints.

GOVERNMENT REGULATION: SIGNIFICANT FEDERAL LEGISLATION

1916   Congress awarded $50,000 to launch an airmail service.

1918   With aircraft and pilots supplied by the army, the U.S. Air Mail Service was introduced. It came under the full control of the U.S. Post Office.

1925   *Contract Mail Act.* Known as the Kelly Act, it was the first major U.S. civil aviation legislation. It took the flying of mail out of the hands of the government, allowed for the airmail contracts to be awarded to private firms, and formed the basis for much of the development of the U.S. domestic air transportation system.

1926   *Air Commerce Act.* Enacted to promote air commerce. The government became the regulator of airlincs through passage of the Kelly Act. It operated and maintained airways, aids to navigation and traffic systems, and provided for licensing of aircraft and pilots.

1930   *Waters Act.* This was the first effort to make airlines self-supporting and to develop passenger markets. Through mergers, Postmaster General Walter Brown forced the formation of United Air Lines, American Airlines, and TWA.

1934   *Air Mail Act.* Also known as the Black-McKellar Act, this legislation separated federal airline authority into three departments, outlawed holding companies that controlled both the operation and manufacturing of aircraft, and, finally, reinstated competitive bidding for airmail contracts.

1938   *Civil Aeronautics Act.* Established three agencies: Civil Aeronautics Authority, Administrator of Aviation, and the Air Safety Board.

1946   *Federal Airport Act.* Provided $520 million to be spent on runways and landing systems over a seven-year period.

1958   *Federal Aviation Act.* Created the Federal Aviation Agency (FAA), one agency to control all aviation matters, both civil and military.

1978   *Airline Deregulation Act.* Intended to reduce regulation and allow free competition in routes and fares among carriers. The act marked the end of 40 years of federal protection and the first time in decades that an entire industry was deregulated. This legislation also called for the elimination of the Civil Aeronautics Board several years later.

CIVIL AERONAUTICS BOARD

The Civil Aeronautics Board (CAB) was an independent regulatory agency. It was established under the Civil Aeronautics Act of 1938, formally put into place by the Federal Aviation Act of 1958, amended in 1968, and declared into sunset (out of existence) in 1985 as part of the Deregulation Act of 1978.

One of the largest single forces on the aviation industry in the United States since it was established, the CAB regulated the economic aspects of the domestic and international air carrier operations. Its area of jurisdiction covered not only route cases but also fares and rates, interchange agreements, aircraft leases, capacity agreements, mergers, ticket discounting, and malpractices. As such, the way that the board conducted its business was the business of Congress. Any changes in CAB responsibilities or functions were legislated by the Congress. Board decisions involving airline domestic operations were not subject to review or approval by the President.

The five CAB members were appointed by the President to a six-year term of office. There could be no more than three members from one political party on the board at one time. This ensured a political balance and a measure of continuity since the tenure of members began with each President's term of office.

The chairman of the CAB was appointed by the President from among the five members. The chairman's term was limited to one year. (Historically, an incumbent President has reappointed the same chairman as long as that person remained a member of the board.)

Although the Deregulation Act ended the CAB's power over routes and fares in 1983 and abolished the board in 1985, the final distribution was subject to change. The future form of the governance of CAB was unclear as responsibilities had to be parceled to other agencies. Most of the agency's functions ended up in the Department of Transportation.

The end of the CAB most certainly does not mean the end of government intervention, for more than 22 departments continue to regulate the air transportation industry in terms of mergers, environment, safety, corporation, consumer action, justice, securities, and so forth.

FEDERAL AVIATION ADMINISTRATION

The Transportation Act of 1966 established a new executive department known as the Department of Transportation. The general welfare, economic growth, stability, and security of the nation pointed to the need for the development of national transportation policies and programs effectively utilizing the country's transportation resources. The act provided for inclusion of the Federal Aviation Agency in the department as the new Federal Aviation Administration (FAA).

Directed by an administrator, who is appointed by the President, by and with the advice and consent of the Senate, the FAA has as its primary function fostering the development and safety of American aviation.

More specifically, the FAA is responsible for developing the major policies necessary to guide the long-range growth of civil aviation; modernizing the air traffic control system; establishing in a single authority the essential management functions necessary to support the common needs of civil and military operations; providing for the most effective and efficient use of the airspace over the United States; and for the rule-making responsibilities relative to these functions.

The FAA constructs, operates, and maintains the National Airspace System and the facilities that are a part of the system; it allocates and regulates the use of the airspace; and, through research and development programs, it provides new systems and equipment for improving utilization of the nation's airspace.

The FAA administers the Federal-Aid Airport Program and provides advisory services to communities in the design and construction of public airports. It also maintains and operates Washington National and Dulles International airports.

Dulles International was the first airport in the world specifically designed for use by commercial transports.

The FAA prescribes and administers rules and regulations concerning personnel competency, aircraft worthiness, and air traffic control. It promotes safety through certification of flight personnel, aircraft, and flight and aircraft maintenance institutions. It reviews the design, structure, and performance of new aircraft to ensure the safety of the flying public.

Because the United States is recognized as one of the world leaders in air transportation, the FAA plays a vital role in international aviation matters. For example, it sends Civil Aviation Assistance Groups abroad to provide technical aid and trains hundreds of foreign nationals every year in Oklahoma City.

The FAA works with International Civil Aviation Organization (ICAO) in establishing worldwide safety and security standards and procedures. Finally, the FAA participates with the National Transportation Safety Board (NTSB) in the investigation of major aircraft accidents.

Chapter 3

The International Air Transport Industry

Courtesy of British Aerospace PLC; British Aerospace Jetstream 41

STRUCTURE

A discussion of the structure of the international air transport industry is particularly complicated for a variety of reasons. It is regulated in an inconsistent manner, the competitive environment is controlled politically to an inordinate degree, and the behavior of the industry is more like a public utility than a profit-seeking transportation system.

After World War II, it seemed that every nation, large or small, strove to develop its own airline that carried the national banner abroad. It was a symbol of the country's prestige. War-surplus transports were available, preferred low-cost loans could be obtained, fuel was cheap, and an upward demand for travel promised continued growth. Technical assistance from an industrialized government could be obtained, and war-trained crews were in abundance.

The industry expanded at a spectacular rate and its technology even faster. The extent of airline service so broadened all over the world that it was soon possible to fly to virtually every national capital and almost any major city with a choice of several airlines.

The effect of this expansion was to increase the economic, social, and diplomatic interactions among nations, stimulate the development and facilitation of investment and trade opportunities, and cause international travel to take a quantum leap in size.

As air transportation became one of the most technologically advanced of the international industries, the less-industrialized countries found it difficult to keep pace with the changes and to maintain competitive proficiency. As the related functions of competition in international travel became more sophisticated, political management caused inefficiencies and national airlines fell behind, lost money, and required subsidies.

Many countries compounded the problem by attempting to modify competitive factors to their advantage rather than addressing the inefficiency issue. The result was further distortion of the already unstable industry structure.

In terms of physical structure, it was common for a nation to designate a "chosen instrument" airline. This was specified as the national flag carrier and was nominally based in the nation's capital to serve international routes. The length and size of the route structure was often in direct proportion to the size of the country by population or geography, although the Netherlands (KLM), Switzerland (Swissair), and Singapore (SIA) are notable exceptions.

If the country were large enough to support a domestic network, there could be one or several domestic carriers, depending upon the air travel needs of the country. The domestic system seeks to satisfy two goals: demand and convenience of the nation's travelers and feeding passengers to the flag carrier at the gateway for international flights.

This is a dichotomy worth noting between regulation as practiced in the United States (prior to deregulation) and regulation as conducted in international air transportation. In the United States, the federal government was detached from the airline operators. The government regulated pricing, capacity, frequency, and routes. The government's overriding concern was with protecting the traveling public, with a secondary goal of maintaining the viability of the system. Government was clearly not involved with the management of the private-sector air carriers. In international air transport, on the other hand, governments are inextricably attached to their flag carriers be they wholly owned, partially owned, or privately held. Thus, governments exercise conflicting roles as regulators of their own enterprises. The political pressures on such a system are quite constraining.

The difficulty with government ownership of its flag carrier (to whatever degree) is that decisions on pricing, entry, flight equipment, and routes are determined by government agencies that must reflect both the political interest and the national interest—but, in fact, representing influence from voters, government officials, critics, and often non-airline imperatives. When government officials are involved in bilateral negotiations, issues frequently emerge as high priority in the national interest while the needs of the airline are shunted to a subordinate position.

A notable example is the case when the United States wanted to build military bases in a particular European country. One of the trade-offs resulted in the flag carrier of that nation being awarded access to more cities in the United States from its capital than all U.S. carriers combined.

British Airways, for example, widely advertises that it serves more American cities from Great Britain than any other airline, U.S. carriers included. Thus, destinations are often bargained away for reasons other than equality among flag carriers, and politics plays an important role. Such political motivations explain why the negotiations for air rights can be so complicated and also why the final bilateral agreement may or may not make a rational compromise of air transportation economic factors.

Violent actions and reactions can cause governments to twist competitive factors to the advantage of its flag carrier. This can be done subtly in a number of ways like limiting frequencies, capacity, fares, times of arrivals or departures, and gate positions. Even though bilateral agreements might award daily service for both carriers, there have been cases where the certification has been held up for one carrier because of insufficient traffic while the chosen instrument airline flew daily. Governments have reached agreement as to numbers of flights, and then one side has suddenly placed a restriction on the number of passengers allowed, a variable that was not discussed in the negotiations.

Another common ploy is that of landing rights, which do not always include the size of aircraft. After negotiations are complete, governments will subsequently limit the allowable size of aircraft in its own interest. An even more insidious caper is to restrict arrival or departure times to off-peak hours, citing lack of gate space as the reason.

One Asian country that shall remain nameless is a case in point. The flag carrier parks at the main terminal entrance in front of the customs office while many of the competing carriers are shunted to tiedowns two to five miles away, a loss in time and necessitating costly bus service.

In one foreign capital, there is a $10,000 landing fee for a Boeing 747 at prime time, more than five times that in the United States. Though this cost is also charged to the government-owned flag carrier, it is merely a paper transaction to take it from one pocket and place it in another. When a South Pacific nation was forced to retract an onerous landing fee increase, it simply imposed a substantial airways usage tax. Many countries now charge an airport head tax (passengers) or a fuel tax. Others assess an airways radar usage tax when overflying the country. All of these and more are cunning ways of imposing economic burdens on competition in a manner that clearly reveals that nations can alter the competitive environment to make penetration difficult for competition and make up for the diseconomies of government management.

Because of the nature of flag-carrier operation, the most important Competitive Variable has been determined to be service. Service is perceived often as representative of the country, a harbinger of things to come for tourists. Pricing has not been involved in most international air transportation strategies because of the stigma of cheap flights and, perhaps, a lack of desire for tourists who are not able to spend money. So international airlines are more service oriented than price sensitive.

In another view, some countries have discovered the wisdom of using flag carriers to bring visitors to their lands and have reduced fares even though this jeopardizes profit. Several years ago Alitalia attempted to reduce fares across the North Atlantic to Rome in the interest of increasing tourist numbers. The problem was that competitors believed that the fares were "loss leaders" and such a strategy, if it were permitted, would devastate pricing across the North Atlantic. The IATA opposed the plan, and it was not long before the pricing scheme was abandoned.

Another important sector of travelers are the economy minded, and many countries have set up non-IATA supplemental airlines to cater to these customers. These carriers have been able to focus on charters and low-cost, no-frills carriage to meet economy demands without degrading the image of the prestigious national carriers. Curiously enough, most major countries now have such supplemental airlines, which compete with the other non-IATA carriers. Although these second-level carriers have the support of their governments, both the protection and the participation are typically less than that reserved for the national flag carrier. The supplemental or nonscheduled airlines have had to survive in a less-regulated, harsher environment. It is interesting to note that history has shown them to be quite successful in their market niche.

This discussion is a simplification of the basic premises of regulation in international air transportation, yet it identifies a structure base that is highly political and irrational, still retaining many of the characteristics of conventional international competition. Perhaps the most disturbing characteristic of the international air transport industry is its inconsistent behavior. Decisions and actions taken may be rational in one instance and bordering on economic insanity in another. Also, governments change and are simply unpredictable, occasionally illogical, and frequently mystifying.

There are many paradoxes that are worthy of discussion. About 80% of the national flag carriers are totally owned or majority owned by the government. The balance of the flag airlines, Swissair and Singapore Air Lines (SIA), for example, are totally privately owned or mostly privately owned.

Whether government-held or not, however, these are the national flag carriers and are protected by government intervention. As expected, the more investment a government has in an airline, the more protection the airline is afforded. As a result, one might expect 100% government-owned airlines to be the most profitable, yet this is not the case. Degree of government ownership (and, hence, protection) seems to be inversely proportional to profitability. Therein lies the paradox.

Two of the most profitable airlines in the world are the privately owned flag carriers of Switzerland (Swissair) and Singapore (SIA). They both defy conventional practices and yet post profits year after year. Swissair is often acclaimed the finest airline, overall, on the North Atlantic corridor, whereas SIA is generally acknowledged as the number-one airline in the Pacific Basin.

As a further comment on government involvement, it is worth

discussing the economic behavior of airlines in Europe, a regulated environment that qualifies as international. In 1982, IATA experts made the following observations on the reasons why European air fares are higher than those in the United States:

- European operating costs are the highest in the world and twice those in the United States.
- European navigation and landing fees are 5 to 10 times higher than in the United States.
- Fuel purchased in Europe is 150% higher than in the United States.
- European sales costs are 2-1/2 times those in the United States because of the complexity of marketing to more than 20 countries with different languages, business practices, cultures, and customs.
- European personnel and total crew expenses are twice those in the United States because of higher social security benefits and more restrictive labor practices.
- Moreover, economies of scale in the United States reflect greater savings in the United States, as European traffic is one-quarter the size of U.S. domestic traffic.
- More than half of European traffic moves on charters (lower yields) against only 5% in the United States.
- Curfews restrict the hours of operation in Europe far beyond that in the United States.

In analyzing the international industry structure, it is apparent that its regulatory behavior is strongly influenced by national and political forces.

Almost every nation has a designated flag carrier. The extent of the route system is generally in relation to the geographical size of the country. It is interesting to note that about 75% of all international routes are two hours or shorter in length (by jet). The internal structure is unstable and behavior highly volatile because of a strong underlying current of national zeal, free-wheeling politics, and intense competition. But what really sets the industry structure of international air transportation apart is the enormity of political participation and resulting erratic scenarios. (Few industries consume as much capital, not many rely as much on large numbers of highly skilled technicians, and probably no other industry involves as many advanced technologies.)

As the international air transport industry regulates itself, it also establishes its own methods of reporting and measurements of performance, which may be curiously overstated or shaded, with supporting evidence gathered from its own sources. Governments exhibit excessive powers to control competition, thus creating an array of obstacles to profitability, hence viability. Reactions are often turbulent and specious.

Many governments perceive their flag carriers as representatives of their power and stature on a world scale; others freely negotiate away airline competitive advantages in the country's larger interests.

The basic industry structure is relatively straightforward with one flag carrier and a supporting internal system. That is where stability stops. International competitive factors are inconsistently weighted and highly variable depending upon governmental intervention. Unpredictability is international air transportation's most pervasive characteristic.

It is not often profit-oriented (though it says it is), it has access to capital (though not in the conventional sense), and its performance targets are notably subjective. Granted, this description is highly impressionistic, but it expresses the differences between international and national structures, as well as the force acting to bring about changes. The truth is, there is an underlying rationality for the industry to function as it does. Political behavior is to be expected when a government runs and reports on its own industry. It is to be understood that the national interest overrides profit goals and that Competitive Variables will be distorted to correct diseconomies when it is within a government's control and interest.

This industry structure has a strong influence in determining the rules of the game as well as the strategic options available to air transport managers. The inconsistencies of Competitive Variables are neither a matter of coincidence nor bad luck. Rather, the competitive environment in which air carriers must function is rooted in an underlying political and economic subsurface that cannot be dealt with as one might expect in a traditional competitive arena.

COMPETITIVE VARIABLES

A dramatic difference exists in the behavior of Competitive Variables in the regulated international environment (where governments maintain tight control and the public sector the prime concern) and the deregulated environment (with little or no controls and profit the objective). The essence of competing in international markets is the arbitrary power of governments, which can issue orders, make laws, and alter Competitive Variables in the interest of its own flag carrier. The government holds the authority to designate who shall serve what routes into its country as well as frequency and pricing.

When government certifies more than one carrier, it typically imposes rigid restrictions on the freedom under which competition can operate. These are hamstrings and controls that governments can arbitrarily impose, which distort and modify the Competitive Variables and the environment in which they function and do not allow for the same cause-and-effect relationships that prevail in a free and unregulated market.

If a government is called upon to give a reason for its actions, it often assumes a protective position of "preventing excessive and uneconomic competition" that, it may be claimed, erodes safety standards, results in inconsistent service levels, and is not responsive to the "public need and convenience." In any case, the two fundamental competitive choices of freedom of entry and independence of competitive action are held in the custody of the government.

The ability to limit, change, or disturb the process of competing in a market renders the use of Competitive Variables in the traditional sense a decision with great uncertainty. To better understand competition in international air transportation, attention should be given to some of the ways and rationale of why and how Competitive Variables are frequently (but not always) manipulated by governments.

1. Fares have different meanings to different countries for political, economic, and social reasons. While one might perceive a need for service in the national interest and relegate fares to a relatively low priority, another government might determine that its airline is a source of employment that far exceeds the need to operate competitively with respect to fares. As a method of bringing foreign currency into the country, the national airline is very effective. Though chosen-instrument airlines typically account for less than 0.5% of a nation's income, they can earn substantial balance-of-payments receipts in other forms of passenger revenue, and price variations are not a major consideration. Even when underpricing, a government can recover additional income from both carriers' passengers by leveling an airport head tax or a tourist visa charge. Clearly, then, pricing airline tickets on the world's routes is a very complex endeavor and can be driven by a number of political, social, and economic pressures.

2. Frequency is the lifeblood of an airline serving a market effectively. Governments can, and often do, reduce the competition's numbers of flights by awarding a limiting number of gate slots. Time of arrival and departure can also be controlled by the device of gate time-slot allocation, making available an arrival slot one hour after its flag carrier arrives (thereby determining its time of departure from the airport of origin) and designating a departure slot one hour after the nation's own airline leaves. There have also been cases where, after a bilateral negotiation has taken place and an agreement reached, one country will start operation with the agreed-to frequency but not allow competition to provide the agreed-to frequency.

3. Market access might not be permitted at all because a government takes an arbitrary position that it can provide adequate service. A country may allow competition to develop a route over a period of time and then place its carrier on the route while restricting competition with the contention that it can better serve the demand. Pan American has recorded a number of route start-ups around the world only to have the national carrier dispossess Pan Am.

4. Equipment is somewhat more difficult to manipulate as the sale of American aircraft is just as important to the United States as the operation of American carriers. The equipment options abroad are limited. Rarely are there more than one or two types of aircraft available and appropriate to serve a market. As a Competitive Variable, there is little differentiation in equipment between various national flag carriers. That is not to say that there has not been some perverting of the use of equipment as a Competitive Variable. A country, for example, might not allow a competitive airline to operate certain equipment on the basis that its facilities will not support the aircraft. This happened in Japan many years ago where foreign airlines were not permitted to operate a particular new model of aircraft. As soon as JAL received this aircraft, however, all were allowed to serve Haneda Airport in Tokyo.

5. Service is the most highly emphasized Competitive Variable in international air transportation. Nations characteristically "pull out the stops" in trying to outdo each other. This occurred in the "Battle of the hors d'oeuvres" across the North Atlantic between SAS and Air France more than 30 years ago, each trying to surpass the other. It was a remarkable battle that escalated to inordinate levels and, finally, went out of control, much to the passengers' delight. Eventually it was arbitrated by IATA and an accord reached as to the number of calories that could be offered in an hors d'oeuvres service.

6. Advertising, publicity, and public contact can be at whatever budget a country might deem necessary. Whereas a profit-seeking nonflag carrier might invest about 3% of its gross revenue in advertising, a chosen-instrument airline could have a much larger budget, at the discretion of the government. Thus, the advertising variable might have an economic or political base for determining its budget.

In general, disputes as to the excessive exploitation and unfair distortions of Competitive Variables in international air transportation are arbitrated by the IATA, if both nations are members. That process is more of an open forum than a jurisdictional situation, for airlines may choose not to abide by the decision of the arbitration. Other disputes are a matter of bilateral discussion and power politics.

Competitive Variables are critical elements in the development and success of an international airline. The sovereignty of airspace permits nations to alter their effectiveness. A country's action in the matter of controlling Competitive Variables arises from its perception of the public need, its possession of air rights, and the responsibility of owning and protecting the flag carrier. Where governments permit Competitive Variables to be used freely, their behavior is similar to the deregulated environment. In most cases, the superior right to shade Competitive Variables to give its national carrier an advantage is too attractive to ignore. To understand the factors involved in the unpredictability of the behavior of Competitive Variables in the international scene, one must examine some underlying influences.

- **Economic**
  Balance of payment, workers in the industry, the nation's economy, tourism, and commerce all affect how a country will deal with Competitive Variables in air transportation.
- **Political**
  This is the force acting on a government in power; it often results in party-oriented courses of action and may twist Competitive Variables to lopsided, totally unfair advantages.
- **Social**
  The national good is a dominant, self-determined factor and rises above rational business behavior. The image of a nation, it is often believed, rides on the wings of its flag carrier.
- **Freedom of entry**
  The ability of a country to have absolute control of airlines serving its capital in pricing, frequency, and capacity is an awesome power.
- **Independence of competitive action**
  When nations can dictate the limits of competitive choices, they have the ability to reduce Competitive Variables to a state of impo-

tence. Where competitive action is contorted, it is impossible to determine what the result will be of a competitive decision.

Competitive Variables play an important role in international air transportation, for they are often uncontrolled or not tightly restrained. When this happens, the behavior of variables is generally straightforward, but when subject to the political whims of a government they are unpredictable and the response to competitive decisions can be so irrational as to be economically dysfunctional.

OPERATIONAL CHARACTERISTICS

Airports

With rare exceptions, most public airports are owned by national governments. In effect, they represent a pass-through subsidy to their chosen-instrument airlines in various ways. The landing fee paid by the national carrier is paid to itself as is the case with fuel tax, airport rental space, and other charges that are assessed on operating airlines.

These fees are at the discretion of the government and vary from country to country. These charges represent a substantial source of income to governments as they are usually structured at fully allocated costs. Moreover, terminal access is rarely doled out equally; rather it adheres to the competitive advantage of the national airline.

The value of tourism has been recognized as being able to change a negative balance of trade to a positive one and countries have committed to great expense to provide up-to-date mammoth airports.

Terminal facilities are of importance to airlines since these facilities control the speed and convenience with which passengers can be served.

Airlines normally lease sales and administrative space at the terminal and in other suitable locations in the surrounding communities. Ground-support equipment and other physical facilities represent about 20% of an airline's investment in physical assets. Of this 20%, about half is accounted for by land and buildings while the remainder is usually represented by ground-support equipment. Because of this high capital investment, ground-service companies have emerged that will provide ground services for a fee. This permits a carrier to operate fewer trips than would support an investment in physical facilities and allows flexibility to increase or decrease flight frequencies.

The airport runways control—by their length, number, orientation, and weight-bearing ability—the volume of traffic and types of aircraft that can utilize the field. When the jumbo jets were ordered, for example, there were only a limited number of airports in the world that had runways long enough to permit their operation, and, in most places, no gates or ground facilities were available even if landing was possible. As discussed in Chapter 2, this resulted in a massive program of airport expansion at almost every major city in the world. Construction is an ongoing process at major airports, and an international pilot was heard to lament that he had a dream "to land at just one airport somewhere around the world which did not have construction taking place."

Air navigation, area radar, and instrument-landing facilities at most major airports are provided and operated by an agency of that government. The International Civil Aviation Organization (ICAO) inspects and monitors the facilities to ensure accepted levels of safety and a degree of uniformity. However, each country "runs its own show" and the range of quality is broad. In Europe and North America, there is a high degree of safety and uniformity. In Southeast Asia and South America, this is not always the case.

By contrast, as mentioned before, the fees or charges to international airlines for using most airports are determined by the host country. The result is a wide variety of methods and levels of charges. The government determines what these fees will be, and each airline works out its own arrangements with the airport administration concerning hangar rentals, counter space, ramps, gate slots, and special incentives to establish or expand facilities. At an increasing number of airports, revenues from these airline fees are a growing part of the total revenue. Concession fees, counter and office space, and other ground leases are among the other sources of income being generally developed. The majority of foreign airports are designed to be self-supporting.

Many countries of the world have one "gateway" airport designed as the country's major airport of entry or exit. In the most highly industrialized nations, there are often more. The gateway airport is likely to be viewed by governments as the window of their country for those landing to stay or pass through. Pride of ownership is a strong and motivating force. It is common to find gateway airports with facilities large and in excess of needs, designed to be a model of the nation.

Equipment

The most important physical assets that an airline owns (or leases) are its aircraft, which account for about 80% of a carrier's fixed assets. Of this 80%, one quarter or less represents spare parts and extra engines. Not only does flight equipment represent most of the carrier's investment but it also dictates operating procedures, determines the airline's image and target market, moderates passenger acceptance, and highlights profit-and-loss potentials. Thus, it may be said that decisions to purchase new aircraft are the most important ever made by an airline's management. Again, governments tend to buy excess capacity aircraft in expectation of traffic growth.

Aircraft are purchased in the United States, France, and Great Britain. Low interest rates, which are supported and underwritten by governments, make aircraft attractive to buy. Export-import bank loans are available to foreign carriers buying U.S. aircraft at rates more favorable than for a U.S. carrier. When Pan Am considered buying the Lockheed L-1011, the decision was cast when the British government offered to collateralize Pan Am's loan if British Rolls-Royce engines were used as the power plant (representing about 50% of the price of the aircraft).

Other than the French Airbus, foreign manufacturers are concentrating on the lower range of commuter and regional aircraft because of the enormous start-up costs involved in the manufacture of airliners. In the Boeing case, the entire resources of the firm were gambled on the new model 757 and 767 aircraft. Foreign manufacturers have also found a viable niche in manufacturing components for American aircraft. The United States clearly dominates aircraft manufacturing and, in almost all cases, at least 40% of the components of foreign-made aircraft are U.S. derivatives under license. Presently, however, the tide is changing, and international manufacturers are increasing their market share of new aircraft.

The dominant engine built outside the United States is by Rolls-Royce; it is excellent in design, more complicated than U.S. engines, and slightly more expensive to operate. Other than the British, no foreign countries have been able to compete with Pratt & Whitney and General Electric in engine manufacture.

Most avionics and support equipment is also provided by U.S. manufacturers or countries producing under license. The same is true of ground-support equipment.

Because governments consider it crucial for a flag carrier to have equipment that is at least as good as that of its competitors, the decision by one nation to purchase a newer, faster, or more comfortable aircraft places pressures upon its competitors over the same routes to acquire the same equipment.

Thus, not only must the economics, operating characteristics, service requirements, and suitability of the new plane be considered, but the action that another nation has taken or will take must also be evaluated. When a major new aircraft is designed by a manufacturer it is well known throughout the airline industry, so one national carrier cannot really surprise another. Years ago, being the first on a route with an aircraft gave a competitive edge, which resulted in an economic advantage. Today, there is a body of evidence that being first may not be advantageous. It has been said that Pan Am spent $17 million on the B-747 in research and development, which later purchasers enjoyed at no cost. For example, the Boeing 747SP (short-bodied, long-range, Special Performance jumbo jet) flew transcontinental for six months on suboptimal flights to prove the aircraft at an estimated cost to the industry of $18 million. Some airlines, Cathay Pacific and others, were among those who did not want to be first.

The Boeing 707 was incredibly successful when it was introduced on scheduled airline routes, because the U.S. Air Force had flight-tested the KC-135 derivative for four years.

Operating characteristics in terms of various airport and airway facilities are positioned widely on the spectrum of reliability. In industrialized countries and high-density routes, the operating characteristics and reliability of facilities are sound and relatively consistent. In nonindustrialized areas and less-traveled routes, operating characteristics are often marginal or poor. There is a wide disparity in airport and airway facilities over the world, and the integrity of individual systems is difficult to control. Gateway airports are typically government owned, in continuous construction, and usually exceeding traffic needs.

Ever since World War II, the United States has dominated in manufacturing transport aircraft. Other countries have participated by building components for U.S. transports and, over the years, have developed expertise. As a result, Great Britain, France, and Germany have focused on smaller transport aircraft with considerable success. Of those aircraft manufactured overseas, about 40% of the components are American made.

Bilateral aviation negotiations between the United States and other nations are influenced by more than equal treatment, for the benefits gained by the United States from the sale of flight equipment may exceed the loss of traffic rights for American carriers. Aircraft, parts, and support equipment represent the largest export dollar of the United States.

Air service, the level of industrialization, national weather, and, ultimately, the air transport sophistication of a country strongly influence its operating characteristics.

GOVERNMENT REGULATION: FIVE FREEDOMS

The expression "Five Freedoms of the Air" is most often used in the negotiations of airline traffic rights and bilateral agreements. In practice, the airspace over a country is far from free. International aviation functions on the principle that every sovereign state has unlimited supremacy over, and control of, the air above and within its borders.

When the international air traffic conference convened in Chicago in 1944 to consider the postwar international air traffic policy under the ICAO, it became apparent that some clarification of air rights was necessary. The Five Freedoms of the Air have become the basic mechanism for identifying all rights under international air agreements.

The first freedom, the right to fly over another nation without landing, is freely granted, except in the protection of military installations or restricted areas. In these cases, restricted airspaces are designated and air corridors described.

The second freedom, the right to make a technical landing for refueling, etc., without picking up or letting off revenue traffic, is usually allowed in the case of emergencies or unanticipated stops.

The third and fourth freedoms are those most frequently the subject of international agreements and are often negotiated together. The right to carry revenue traffic from the carrier's base nation to a treaty partner's territory, and the right to carry traffic from the partner's territory to the carrier's base nation, comprise the definitions of these two freedoms.

The fifth freedom, the right to carry revenue traffic between any points of landing on flights among three or more treaty partner nations, is the most sought after because it can provide extra traffic to support low-density routes.

In addition to the five basic freedoms, three supplementary freedoms were identified and acknowledged to be of minor importance but worth describing. These include: the right to carry revenue traffic between two treaty partner nations through carrier's base nation (sixth freedom), the right to fly revenue traffic between two nations by carrier of a third nation (seventh freedom), and the right to carry revenue traffic between two points within a foreign nation (eighth freedom, or cabotage).

Most airlines operate internationally under limitations imposed by bilateral agreements negotiated by their respective governments. Bargaining for traffic (capacity) and landing rights (frequency) takes many forms. The most common formula was developed by the United States and the United Kingdom in the historic Bermuda Agreement of

1946. At this meeting, the two major airline nations of the time established a fair basis for traffic and landing rights, based on its nationals flying on the route, without abuse by either side.

INTERNATIONAL AIR TRANSPORT ASSOCIATION (IATA)

The IATA is the international organization of more than 110 scheduled airlines. Its members carry the majority of the world's international air traffic under the registration of more than 80 nations.

The IATA is dedicated to the coordination of the world's scheduled service. Its primary function is to motivate the movement of traffic with the greatest possible speed, convenience, efficiency and economy.

Some have suggested that the role of the IATA is similar to the Civil Aeronautics Board on an international basis or have likened it to an aviation United Nations. The association was formed in 1919 and reorganized in 1945. Headquarters for the IATA are located in Geneva, Switzerland.

Membership in the world air body is voluntary. Any operating airline that has been certified by its government is eligible for membership. Full and active membership is accorded to air carriers engaged directly in international operations, whereas domestic airlines are associate members.

IATA Aims and Services

Once governments have promulgated a formal exchange of traffic, capacity, frequency, and other rights through bilateral agreement and have licensed the airlines selected to perform the service, IATA's work begins. Ever since its formation in 1945, the association has established the following three specific aims:

To promote safe, regular, and economical air transport for the benefit of the people of the world: to foster air commerce and to study the problems connected with it.

To provide means for collaboration among air transport enterprises engaged directly or indirectly in international air transport service.

To cooperate with the International Civil Aviation Organization (ICAO) and other international bodies. Throughout its history, the IATA has been closely associated with the

ICAO, which creates world standards for the technical regulation of civil aviation.

Underlying Purpose

For the airlines, the IATA provides the machinery for finding joint solutions to problems that cannot be adequately dealt with by government negotiations. The IATA has emerged as the mechanism by which the airlines have woven individual route and traffic mediation procedures into a workable worldwide public service system, despite the differences in laws, languages, political imperatives, currencies, and measurements. The association has developed a pool of experience and information, and it oversees many common services and activities. Thus, the IATA, the collective personality of most of the flag carriers of the world, functions as the international air transport industry's linkage with governments and with the traveling public.

In addition, the association furnishes governments with the medium for negotiating international rates and fares agreements. It strives to work out the rapid and economical transport of airmail, and to ensure that the interests of commerce, convenience, and the safety of the public are served.

For the benefit of the traveling public, the IATA attempts to establish efficient operations and business practices by airlines and their agents, freedom from regulatory constraints, and the lowest possible fares and rates. Passengers can fly almost anywhere in the world on tickets or waybills of one single IATA airline. Revenue exchanges resulting from one airline's acceptance of another's documents are settled centrally among the airlines, much the same as when banks exchange checks. If the airlines were to handle these transactions themselves, the exchange would require an enormous organization. Instead, the entire task goes through the hands of relatively few people in the IATA's clearing house by offsetting members' accounts.

Internal Organization

The IATA's annual general meeting provides the basic source of authority for the organization. All active members have one vote. Year-round policy direction is determined by an elected executive committee. Its creative work is largely carried out by financial, legal, technical, traffic advisory, and medical committees. Traffic conferences establish the negotiating forum for fares and rate agreements that are valid for periods of two years.

Administration of the IATA is headed by a director general and three assistant directors-general for each of traffic, technical, and general counsel. A second main association office is in Montreal, while regional technical representatives are based in Bangkok, Nairobi, and Rio de Janeiro. Traffic-service offices are located in New York and Singapore.

The IATA's budget is financed from the dues paid by its more than 100 members, largely in proportion to the part of the total international air traffic carried by each airline. Some IATA activities are self-supporting through charges for services rendered.

It would not be inaccurate to refer to IATA as the world parliament of international air transportation. The association has two objectives; its operational objective is to ensure that the aircraft used to carry the world's passengers and goods are able to do so with maximum safety and efficiency, under common and defined regulations; its commercial purpose is to ensure that passengers, cargo, and mail can move anywhere in the world without constraint.

As a result of airline cooperation through the IATA, an individual passenger can, with a single telephone call and payment in a single currency, arrange a journey that will include many different countries and the systems of several scheduled carriers.

Both the operational and commercial categories of IATA activities are concerned with the cost of airline operations and the carriers' charges to the public. The aim, specifically, is to keep both as low as possible while maintaining high standards of safety and service. The association makes a constant effort to simplify and standardize services, procedures, and documentation among airlines, among governments, with manufacturers, and in collaboration with other international organizations.

INTERNATIONAL CIVIL AVIATION ORGANIZATION (ICAO)

The ICAO was established by more than 130 sovereign states in 1944 to serve as the communicative medium through which international understanding and agreement could be reached on the more technical aspects of air transportation to assure quality control and commonality of procedures. Additionally, the ICAO is dedicated to developing the principles and techniques of international air navigation and to foster the planning and development of international civil aviation.

ICAO's Aims and Services

To ensure   the safe and orderly growth of international civil aviation throughout the world.

To encourage   the arts of aircraft design and operation for peaceful purposes.

To encourage   the development of airways, airports, and air-navigation facilities for international civil aviation.

To meet the needs of the peoples of the world for safe, regular, efficient, and economical air transportation.

To prevent   economic waste caused by unreasonable competition.

To ensure   that the rights of contracting states are fully respected and that every contracting state has a fair opportunity to operate international airlines.

To avoid   discrimination among contracting states.

To promote   safety for flight in international air navigation.

To promote   the general development of all aspects of international civil aeronautics.

ICAO's Underlying Purposes

The ICAO works in parallel with the IATA and in cooperation with it on the standardization of navigational aids, questions of air traffic control, ground communications between aviation centers, technical equipment on aircraft and at airports, the assembling of international aviation charts, and compilation of traffic and accident statistics. The ICAO defines standards and issues recommendations that are often adopted by aviation authorities and airlines all over the world. Technical assistance and schools that the ICAO sponsors in developing countries help promote the development of commercial aviation in those lands.

Internal Organization

The sovereign body of the ICAO is the assembly. The governing body is the council. The assembly meets at least once every three years and is convened by the council. Each contracting state is entitled to one vote. Decisions of the assembly are taken by a majority of the votes cast except when otherwise provided in the convention. At this session the complete work of the ICAO in the economic, legal, and technical assistance fields is reviewed in detail. Guidance is given to the other bodies within the ICAO for their future work. The council is a permanent body responsible to the assembly for a three-year term. One of the major duties of the council is to adopt international standards and to recommend practices.

Although the ICAO may become involved in many matters as an arbiter, in general it works toward the maintenance of safety and regularity of operation of international air transport.

A secretary-general heads the secretariat, which is divided into five main divisions: the Air Navigation Bureau, the Air Transport Bureau, the Technical Assistance Bureau, the Legal Bureau, and the Bureau of Administration and Services.

Activities of the ICAO are more technical in nature than those of the IATA which are more procedural. A parallel might be drawn between the FAA and the ICAO with the CAB and the IATA. In general, the ICAO is concerned with the following areas:

1. *Standardization,* which is the establishment of international standards, recommended practices, and procedures covering the technical fields of air transportation.
2. *Regional planning,* which is a function derived because some problems cannot be dealt with on a worldwide basis. The ICAO has identified nine geographical regions that must be treated individually for planning the provision of air-navigation facilities and services required on the ground by aircraft flying in these regions.
3. *Facilitation,* which is the effort by the ICAO to ease the impediments placed by customs, immigration, public health, currency control, and other formalities on the free and unimpeded flow of passengers and cargo across international boundaries.
4. *Economics,* which relates to the statistical sources that the ICAO collects and publishes and upon which member states and airlines can base their future plans.
5. *Technical assistance for development,* which is the special attention that can be given jointly to promote air transportation in developing countries in all phases of the industry.
6. *Law,* which is the attempt to unify influence for the development of a code of international air law. Legal philosophies and systems of juriprudence are combined into a form that is acceptable to all member states. This process is, of course, continuing.

PART 2

Description of the Air Transport Industry

Chapter 4

Organizations and Management

Courtesy of Airbus Industrie of North America; Airbus A-320

/

structure and management are intricately related and
rmine to a large extent the success of an airline. (Air
anizations have emerged primarily because experience
showed .... in a complex environment, organized groups pursue
goals and objectives more efficiently than do isolated individuals.
However, these organizations have evolved to be more than mere instruments providing air transportation decisions.) They also exert a profound influence on those decisions. Additionally, they provide a
framework within which management can coordinate their efforts toward a common goal—to secure a position in the marketplace allowing the carrier to compete successfully and achieve a profit.

In fact, one can say that (the most important characteristic of airline organizations is their ability to pursue goals efficiently and effectively. Two main factors determine their success: (1) the intelligence of
the organizational structure; and (2) the intelligence of the organization's management. To meet the challenges of a complex competitive environment, an organization must be well designed and well
managed. \

HISTORY

Modern airline management is striving to develop an organizational
structure with a profitable distribution of attention between operations
and marketing. The importance of marketing has really only become
evident since deregulation of the industry in 1978. Before that, government regulation and subsidy created an artificial environment that effectively constrained the options available in a free market.

When airlines were first formed, the primary concerns were the
technical aspects of operation. Although a market was recognized inas-
much as American air transport firms carried mail and European airlines carried passengers, technological limitations mandated the
structure of the airline organization. In most cases, the key officers
were pilots or technicians. Technology restricted the first airlines to
short, slow flights, always contingent upon good weather conditions.
Maintenance was excessively high and experimental. Most decisions
had a strong technical bias, and although the potential of the market
was recognized early on, the actual market was quite small.

World War II brought a much larger market into play, so to speak,
with all its attendant complex logistical problems. Meanwhile, technology had advanced considerably, giving airlines much more flexibility.

The two conditions combined to complicate airline operations: Formal organizational techniques and scientific management became necessary to meet the needs of the growing industry. Still, operations were far more important than marketing, for the war assured an ebullient market and government subsidies protected the industry from dangers of economic losses. During the war, the attention of airline organizations was divided between technological and managerial aspects of operations.

Fortunately, the newly important administrators were able to benefit from the scientific management philosophy of Frederick Taylor, which had been, for the most part, ignored in the industry since the inception of air transportation 50 years earlier.

The post–World War II boom in aviation saw the creation of a remarkable array of new airlines due to the availability of war surplus aircraft. A pioneering and entrepreneurial spirit prevailed, and the founders of these airlines dominated their organizations and exercised unilateral powers. Individual personalities determined the structure of the organization.

For example, C.R. Smith of American Airlines focused on customer convenience and service. Bob Six of Continental Airlines was a buccaneer and renegade in the industry; his airline was aggressive, competitive, unpredictable, and dedicated to expansion. C.E. Woolman of Delta Airlines emphasized employee participation, paternalism, and nonunionism. Eddie Rickenbacker of Eastern Airlines centered on efficiency and economy. Donald Nyrop of Northwest Airlines was the most economy-conscious operator of all; his firm was known to be the lowest cost operator in the nation. Juan Trippe of Pan American World Airways envisioned his airline as the international flag carrier of the United States. Jack Frye of TWA strove to be the first with the newest and most advanced equipment, a view shared by the majority stockholder, Howard Hughes. W.A. Patterson of United Air Lines ran a conservative organization, reflecting his sound understanding of finance.

The competitive environment of the air transport industry had also changed dramatically. The markets burgeoned as passenger traffic was added to mail carriage in the United States, and added conveniences and comfort attracted a larger number of passengers throughout the world. Moreover, the range of aircraft increased to add market pairs that had never before been feasible. The management techniques and tools put to use to maximize logistical support during the war continued to serve airline organizations during the 1950s.

More and more, what distinguished the different organizations was not technology or supply but rather management. As the industry grew worldwide, increasing organizational demands made single-personality dominance impractical. Solo seat-of-the-pants decision makers gave way to corporate team management as the business became more sophisticated. The scope of management shifted to meeting changing public needs, technological control, and legal restrictions. In particular, though, the demands of the marketplace, especially those of the passengers, took on an unprecedented importance. New forms of organizations evolved to fill the requirements of the new era. The matrix organization dominated the industry because it allowed for decentralization of the marketing structure. Decentralization was necessary in the face of increasing size of markets, dynamics, complexity of market variables, and delegation of both authority and responsibility.

The 1960s brought the flowering of the Jet Age; now no two places on earth were more than 24 hours distant from one another. With this new era came rapid increases in cost of operations and intensity of competition. The various organizational structures continued to fall into a pattern reflecting a choice of emphasis: operations (technology and organization) or marketing (service, passengers, and frequency). Each organizational structure has its own set of advantages, but neither embraces the advantages of both structures. If the operations structure is dominant, an increased integrity of supply results but only with added costs and decreased revenues. If marketing is dominant, capacity and revenues increase, but operating costs rise measurably. (See Figure 4.1)

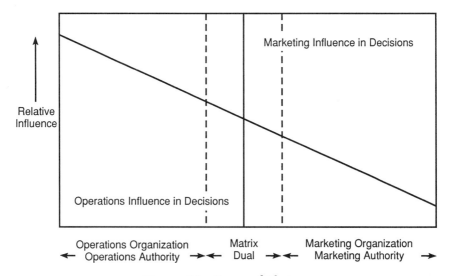

Figure 4.1   Range of Choices

Thus, the 1960s demonstrated the need for a balance between operations and marketing. Management training, equipment and maintenance planning, inventory control, strategic planning, productivity, purchasing, pricing, and utilization are some of the areas of specialization that emanated from managerial attempts to meet this need.

Management throughout the industry recognized the importance of optimizing resources in a capital-intensive environment. Subsidies were no longer available to the trunk carriers, yet government regulations were so onerous that few airlines were more than marginally profitable. As unions developed and competition increased, organizations struggled to meet the demands of markets while maintaining control of operational factors.

Airlines abroad continued to be heavily subsidized and, for the most part, owned by their governments. In general, traffic and financial reporting was controlled and various methods used. Competitive variables were often the subject of bilateral negotiations. It was about this time that U.S. airlines appeared to follow a different strategy than did carriers abroad.

Most of the pioneer leaders of the industry had risen to corporate levels or emeritus. Their influence was generally out of the mainstream of daily operations and this seemed to be a worldwide phenomenon. Management moved more toward a science than an art and was committed to formal organizational practices and management science techniques. Five functions of management proved critical: planning, organizing, directing, controlling, and coordinating.

The 1970s catapulted the industry into both the wide-bodied era and deregulation with an added gigantic increase in fuel prices. The result was broad and spurious changes and experimentation. With deregulation, marketing rose to the superior position and had a significant effect throughout the world. This caused another permutation of organizational structures.

STRUCTURE

Airline organizations assume a variety of forms. Generally, they are either government, publicly, or privately owned. Thus the organization may be a single proprietorship, a partnership, a corporation, a conglomerate, or a ward of the state. In size, an airline may range from a commuter line with a few employees to an international flag carrier with tens of thousands of employees. Air carriers may develop service between two or more cities in a line, in a hub-and-spoke configuration, or in a regional grid; they may set up a long-haul operation or serve a worldwide network. There can also be any combination of the above.

Figure 4.2—Frequent Airline Organization Matrix—gives a general idea of the organizational hierarchy of a typical airline company. In fact, this structure is typical of most publicly held corporations in the United States. It should be stressed, however, that the matrix depicted in Figure 4.2 is not complete or entirely accurate. Charts cannot indicate the scope of interactions between departments in such complex organizations; they tend to oversimplify for the sake of clarity.

As competitive, operational, and environmental changes take place, the basic form of the matrix remains stable while emphasis changes to reflect the pressures and beliefs of management. For a variety of reasons, financial or otherwise, planning may be perceived as less important and fall under the finance function (Figure 4.2). Flight service may be determined to be a marketing function and not an operational activity. Airlines vary in their ability to exploit market instabilities because of the flexibility or inflexibility of management.

Reaction to relatively short-term oscillations in the supply and demand of markets may be immediate or take up to six months or

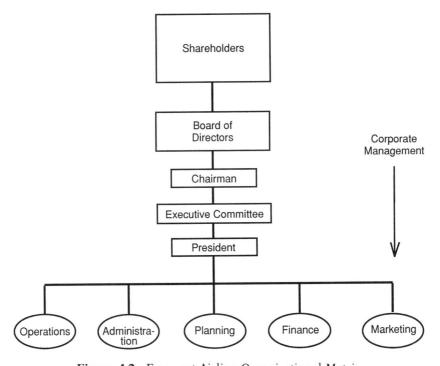

Figure 4.2 Frequent Airline Organizational Matrix

more. The matrix cannot describe the differences in management styles; they will vary from lean in number to large staffs, from relatively stable and slow to act, to dynamic and quick to change teams.

An organizational format attempts to delineate management's form in terms of its position in the competitive environment. For the most part, organizations are structured to deal with two primary forces, which are outlined below.

1. INTERNAL ENVIRONMENT

a. **Macro-environment:** The organization is structured to optimize market opportunities. The organization also has an infrastructure and interactive process to facilitate management actions, plans, and the achievement of goals.

b. **Micro-environment:** The organization has two tasks systems, which consist of base suborganizations dealing with implementation and daily operations.

2. EXTERNAL ENVIRONMENT

a. **Macro-environment:** This segment includes demography, economics, resources, technology, laws, politics, cultures, national sovereignty, and other forces impacting on the organization.

b. **Micro-environment:** This segment includes all of the publics that have a real, potential, or perceived interest in the organizational mechanics. The airline must be concerned with these groups' potential effect on the carrier's ability to achieve its objectives: financial, community relationships, safety, noise abatement, schedule integrity, daily performance, and passenger preference.

Micro-environments (both internal and external) are characterized by possessing a rather dizzying inability to remain static for any predictable period of time. It is by nature a hyperactive form, and rarely can be stationary. The fundamental impulses that set and keep the micro-environments in motion come from new market variables; new consumer demands; new product in terms of aircraft; changes in the economy, and new financial input, all of which can come into existence in a market in a short length of time. They can be of such spontaneity that the organization dealing at the micro levels must be adaptable in meeting daily, even hourly, changes. When environments are uncontrollable they are also unpredictable. Management must be ready to react and realign to spot instabilities. Although the frequency of reactions is high, the magnitude of decisions is not large. Yet management must understand the central planning objectives of the firm, as the tendency is for reactive market decisions to reduce product differentiation. Innovative decisions at a low level seem to maintain a product difference whereas reaction in matching has the opposite effect.

The courses of action in the micro-environment, therefore, must be monitored and evaluated carefully. The macro-environment, on the other hand, is relatively straightforward with only moderate oscillations. From these facts arise the central issues of flexibility, ability, and implementation in all environments.

PROCESS

Organizations have expanded in the current operating environment. Present airline operations require so many different highly developed skills that it is not possible for one person, or even a few, to possess them all. The one-leader management is, almost by definition, a barren one with little interaction, and its success depends upon an uncomplicated industry where quickness of decision making is important. As the industry grew, organizations were compelled to segment responsibilities and establish new procedures.

Top management must clearly define the path of problem-solving with respect to long-term objectives, but must not become involved in dealing with the multiple phases of the industry.

It is difficult for air carriers to achieve and maintain leadership in markets of rapidly changing consumer fancies and innovative competition. The airlines weave themselves ever more deeply into reacting to preserve market share. This type of managerial style always has unforeseeable and possibly dangerous results. In early stages, it may pass unnoticed, but it can become uncertain, inefficient and, if wrong, costly.

The competitive environment was more constant in times of regulation. But the realities are that the more comprehensive the regulatory system, the more surely it will be dominated by mediocrities and the more trivial will be the growth of the industry. Passengers had fewer choices and Competitive Variables were more restricted under regulation. A carrier could survive by serving the same markets, at relatively inelastic pricing tactics, year after year.

The most serious damage inflicted by government controls was the discouragement of innovation and the perpetuation of government-approved obsolescence. It was unfortunately true that the government often gave a distorted idea of the actual market conditions.

Coping with the reaction process in a deregulated market calls for considerable facility.

The organization's culture must be better able to handle the resources available, convert these resources into useful travel products, and allocate them to a variety of competitive choices in an optimal manner. The process must be reasonably disciplined to accommodate changes in personnel as necessary, and responsibility as required.

The organizational process might follow this five-year critical path:

ENVIRON-MENT	OBJECTIVES	STRATEGY	STRUCTURE	SYSTEM
(Market)	(Profit)	(Resources)	(Plan)	(Action)
1990–1995	1992–1995	1990–1995	1990–1995	1990–1995

This is a simulated case of an airline organizational process believed to be sound in 1985. If the process is not flexible and adaptable, the plan process may be firm and efficient but not effective.

If the structure of an air transport organization is to permit maximization of the airline to the market, the process is the actualization of the transport of the highest number of passengers at the lowest possible cost. Finally, structure is doing the right thing, whereas process is doing the things right.

The following anecdote, from *Build a Better You—Starting Now*, by Donald M. Dible clarifies the efficiency vs. effectiveness distinction. A bank messenger is toting a pouch containing five $100 bills and one hundred $1 bills. A gust of wind blows the bag out of his hand and it flies open. The bills are strewn all over the street.

An efficient messenger would start picking up bills from the point where he was standing. An effective one would pick up the $100 bills first. This admittedly exaggerated anecdote illustrates the book's point: It is more important to do the right thing than to do things right.

BEHAVIOR

"The times, they are a-changin,'" Bob Dylan, the strolling minstrel of the Jet Age, sang. With deregulation there has certainly not been airline "business as usual," for the equilibrium of "franchise" and "regulated rate of return" has been abandoned.

These two factors provided a reasonable degree of stability, which permitted longer intervals of time between the type of disruptions that today are so complex, diverse, and spasmodic that an organization has great difficulty maintaining a prolonged balanced behavior. Its micro-organization structure is challenged more often, and its processes are continually made less effective by unforeseen events. This is seductively true in regional carriers, but as airlines grow in size and maturity, they experience increasing difficulty in preserving a rapid response system owing to the nature of large corporations. Confronted with massive and inescapable change, the airline that is constructed on a single-point future is doomed.

The organization must be able to respond to any of a series of plausible scenarios, and must be prepared for such scenarios to become reality. The organization and its environment make up an ecosystem. Disturbances may be perceived as uncertain or positive.

Depending on an organization's view of stress, disruptions will result in opportunities for some and crises for others. The properly "tuned" organization (structure and process) will have a built-in behavioral pattern for identifying, appraising, and reacting to any disruptions arising in its environment.

Behavioral conduct of an organization depends primarily on whether it has a marketing or operating bias. The focus of marketing differs quite widely from that of operations. It follows that, in a simple air carrier, the basic organization model would vary slightly in structure, moderately in process, and considerably in behavior depending upon whether its management is oriented toward marketing or operations.

Matching the airline to supply and demand (operations to marketing) in target markets is the key to survival. Several caveats must be kept in mind. First, the organizational design must be constructed according to its available resources and must properly identify the market position in which the company's strengths can be most effectively exercised. The game plan must be precise yet include an array of plausible alternatives. In the micro-organization, a set of travel concepts and services must be carefully articulated and made available and accessible by selective frequency, capacity, and pricing. The travel offerings must be presented in a rational manner (supply) to the appropriate market (demand), in order to convince the prospective traveler to exchange resources for those benefits. The macro-system deals with market positions, planning, and future scenarios.

The organization must have a processing system that supports the supply/demand strategies. The execution of the process must be adaptable and convertible in the face of market oscillations.

The Competitive Variables constitute a serious problem for organizations because of their inconsistent rates of change. When must the organization react to competitive forces? Response to every competitive action follows a bell curve and creates a trade-off analysis. It has been noted that demand is income elastic. At what income variance does an organization enter or leave a market? Gross national product (GNP) affects total demand, and at what GNP level does market entry make economic sense? It has been observed by many that price elasticity of demand is probably close to unity.

The discussion of air transport organizations and management is debatable and was meant to be. Clearly, however, the design of an organization must be the "right thing." Management, on the other hand, must "do things right" and behave in a responsive and flexible manner. For management, the important mission is the effective execution of goals and objectives of the organization. The best indicator of a management's success is a profitable airline.

LABOR

The airlines (and railroads) are governed by the Railway Labor Act (RLA), which was enacted in 1926 and which requires employers to bargain collectively and not discriminate against their employees for joining a union. The act also provides for the settlement of labor disputes through mediation, voluntary arbitration, and fact-finding boards.

Unionism in the U.S. airline industry began in the 1920s and experienced rapid growth in the 1930s as a result of a number of factors, particularly the passage of the National Labor Relations Act of 1935. This act established the first national labor policy for protecting the rights of workers to organize and to elect their representatives for collective bargaining.

Pilots organized in 1931 as the Air Line Pilots Association (ALPA). Two years later, ALPA took a stand when one airline president, E.L. Cord, attempted to slash pilot pay. The pilots answered with the first commercial air transportation strike, which lasted two months. It indicated the priority of wages in terms of economic settlements because of the still rudimentary state of most collective bargaining and contracts. Also, the post-Depression union goals focused on wages rather than on other benefits. Unions continued to make substantial progress, but before labor and management negotiators were able to become accustomed to their new relationship and develop bargaining skills, World War II brought economic controls, which postponed the use of union strength and restricted gains. These controls caused a breach and contributed to the bitter confrontations that followed the war, confrontations that have resulted in continued turbulence in labor relations.

Since that time, a number of strikes have hit the airline industry, and individual airlines have had to deal with eight, nine, or more different unions. The Air Line Pilots Association has a reputation for being one of the strongest and best-organized unions in the United States.

The period since 1948 has been one of the most dynamic eras in the history of the American labor movement. Labor disputes have been bitter, and labor gains in the airlines have exceeded those in other industries; at the same time, productivity has declined at a rate higher than other industries.

Several crises have contributed to a radical change in labor's wage/productivity curves; both the fuel increases and the recession of the 1980s compelled airline unions to take wage cuts and to provide productivity increases. The future for labor is clouded; it would appear that ground will be lost for the movement until the industry achieves a reasonable return on investment.

The basic federal legislation providing for labor relations and collective bargaining continues to be the Railway Labor Act. Under this act, disagreements are to be settled by conference. When conferences fail, a National Mediation Board (NMB) is provided to hear grievances. This board is not an enforcement agency, and it cannot compel either unions or management to do anything. If the NMB is unable to resolve the differences between the negotiating parties, it might persuade the parties to submit the case to arbitration. If either party refuses to arbitrate, the unions must wait at least 30 days before striking.

If a strike is threatened and the President of the United States determines that a substantial interruption to interstate commerce is likely to result, the President may appoint an emergency board to hear the dispute. If such a board is appointed, no strike may be called until 30 days after the board issues its report. With all these delays, the negotiating parties may go a long time in postponing the crisis point. (See airline collective bargaining procedures under the Railway Labor Act.)

The threat of a strike to the airlines is a particularly onerous one. Unlike firms in other industries, the demand for air transportation service generally does not backlog, nor is it possible to inventory in anticipation. Accordingly, if there is a strike, the business lost during that period is lost forever.

To offset losses from one carrier to another when a particular airline was struck, six major airlines evolved a Mutual Aid Pact in the late 1950s. This pact was filed with and approved by the Civil Aeronautics Board. It provided for a method that would pass to the airlines suffering from a strike the additional income accrued to the other carriers by reason of diversion of traffic from the struck carrier. It also guaranteed certain basic operating costs. The pact seemed to be suspiciously beneficial to certain carriers.

It was remarkable that Northwest Airlines appeared to have developed termination dates of their labor contracts to coincide with low revenue periods. Some pact executives complained that Northwest allowed more strikes than any other major airline, and these occurred during the low travel periods when the company could be protected by the Mutual Aid Pact. (The pact was abolished as part of the deregulation process in 1978.)

In the 1950s, a bitter battle occurred between the ALPA and the Flight Engineers International Association (FEIA). The issue arose over the makeup of cockpit crews for the new jets. Prior to the jets, most airline crews on four-engine aircraft consisted of two pilots and a third crew member designated as the flight engineer officer (FEO). The FEO was concerned with propeller and fuel mixture controls, cowl flaps, engine superchargers, fuel management, oil cooler flaps, and other mechanical applications. These functions were done electronically, by and large, in the jets or else disappeared, and the only functions left for the FEO were fuel management, preflight inspections, and the management of air conditioning and heating systems.

On the other hand, the workloads of the pilots were much more complicated on the jet aircraft. The ALPA then took the position that the third crew member in the jet cockpit should be a pilot or at least have pilot training, so that this person could take over in emergencies. The FEIA view was that regardless of the number of pilots in the cockpit, there should be at least one professional flight engineer aboard (who did not have to be a pilot and who would be a member of FEIA). This would provide jobs for non-pilot members of FEIA. Another reported goal was to avoid absorption of the FEIA by ALPA.

The airlines were caught in a situation in which the engineers threatened to strike unless the airline guaranteed them a job on the jets, while the pilots threatened to strike if the airlines did guarantee the engineers a job. Negotiations took a different course with each airline and, by 1965, the question was on its way to settlement.

Today's airline flight monthly pay formula for pilots allows for computation in two parts: the four elements making up basic pay (base pay, hourly flight pay, mileage pay, and gross weight pay); and work rules that produce all of the other elements of pay such as duty time protection and trip-ratio pay. Pilots also enjoy fringe benefits and paid-up pension plans. Fringe benefits can equal almost 45% of a pilot's salary in travel, medical, dental, sick leave, and disability retirement. Paid-up pension plans typically provide from 40% to 60% of the pilot's wages over the previous five years as a retirement income at age 60, the mandatory age.

UNION ORGANIZATIONS

Generally, the airline unions are organized by crafts and classes, the two major exceptions being the International Association of Machinists (IAM) and the Transport Workers Union (TWU). The IAM represents various crafts and classes and is especially strong among the mechanics as well as the stock and stores employees, less so among clerical and related employees, and even less among the flight crews (flight service, engineers, and pilots). The TWU also represents employees in several diverse classes as flight dispatchers, flight service, some mechanics, clerical and related personnel, as well as stock and stores people.

Historically, unions have attempted to organize the industry and all have been unsuccessful. In the late 1940s, ALPA began to organize all flight personnel. Within a generation, many of the ALPA affiliates had been absorbed by other national or local single-class unions. Currently, the NMB distinguishes between about nine major employee representation classifications. The majority of these crafts or classes are within its particular union. Some fragmented single-cell unions have formed, but such a union lacks the strength of a national group and few have succeeded.

With all of the collective-bargaining agreements reaching maturity, and with negotiation conferences that can last for months, contractual expirations, and jurisdictional and representational disputes that carriers experience, it is little wonder that harmonious industrial relations prevail in the air transportation industry only by exception, for the ambience is most often a constant state of conflict between labor and management.

MERGERS AND ACQUISITIONS

Most airline growth occurs through internal expansion, which comes as a result of a carrier increasing its share of a market, fleet size, total capacity, or entering new markets. The most spectacular growth and often the largest increases in the airline's total assets occur when a merger takes place.

The primary motivation for airline mergers is to increase the economic performance of the combined carriers. If two airlines merge and the combined firm's value and/or earnings exceed that of the individual firms, then synergy is said to exist, and such a merger is to the benefit of the shareholders of both airlines.

Merger synergism can emerge from three sources:

1. Economies of scale (operations and marketing)
2. Financial strength (ability to survive business cycles, competition, and lower cost of debt)
3. Market strength (reduced competition or stronger ability to perform in markets)

During regulation, prior to 1978, though economies of scale and financial strength were socially desirable, mergers that might reduce competition were not viewed as being in the public interest of need and convenience and were rarely permitted. Most often, mergers took place to avoid a bankruptcy, which could result in the discontinuation or reduction of service. The inordinate length of time required to consummate a merger, such as wallowing through governmental hearings, made expansion through merger an unattractive strategy. Added to that were the time factor, the uncertainty of government approval, and the legal expenses involved.

The chances are high that a merger was rarely considered seriously. After the industry was deregulated, however, all three factors—time, uncertainty, and legal costs—were reduced and mergers began to take place.

There are four generic types of airline mergers:

1. *Horizontal* (two airlines of the same category merge, such as Pan Am and National, United and Capitol, Northwest and Republic, etc.)
2. *Vertical* (an airline merges with a non-airline firm, which is in the production stream such as an airline limousine service, a ground-servicing company, or an airport fueling concessionaire)
3. *Congeneric* (a merger or acquisition with related but not horizontal or vertical, such as a hotel chain or car-leasing firm; United and Westin Hotels, PSA and Valcar)
4. *Conglomerate* (unrelated enterprises combine: TWA acquired Century 21 Realtors, Hardee Restaurants, Hilton Hotels International, etc.)

Economies of scale and market strength are at least partially dependent upon the type of mergers involved. Horizontal and vertical mergers more often result in the greatest operating benefits and, to some degree, competitive synergies. It is probably true that a merger of parallel carriers (grid pattern) has less synergy than does an inline merger (where one carrier's routes are added to another). When routes added to each other traverse enormous lengths, each firm is often permitted to function separately as a profit center division. The congeneric merger, where the traveler can be sold an entire package, offers considerable synergies. Conglomerate mergers characteristically reduce the amplitude of cycles of business as they are in unrelated businesses. Such firms are generally operated as separate profit centers. In any event, airline mergers must be synergistic to succeed, and that is management's most difficult task.

In the majority of merger occurrences, one carrier (usually the weaker of the two) is acquired by another airline. Rarely will the acquired carrier initiate the merger action, unless it is to avoid failure. This relationship is generally identified as the seeking-to-acquire by the Acquiring airline, and the one that it seeks to acquire as the Target airline. The process is relatively simple: The Acquiring airline determines how much synergy will occur, its strategic needs, and the value of the Target airline. Once this has been completed, the negotiating process takes place.

The final result, from a financial point of view, is an Operating merger, in which the two airlines are joined as one, with the expectation of synergistic benefits; or an Economic merger, in which each will evolve into profit centers and from which less economies of scale are expected. Fundamental perceptions are that an Operating merger falls into a short-term strategy with anticipated higher profits and an economic merger should provided a more stable business cycle in the long term. The outcome of parallel carrier mergers might fall into the Operating category, whereas a merger of inline carriers represents Economic mergers in nature.

Having discussed the merger process, it is important to review U.S. merger history briefly. (International carriers are government owned or controlled, for the most part, and rarely merge.) During regulation, public need and convenience were dominating forces. If an airline met with financial difficulties, the problem could be met by government subsidy, low-cost loans, route awards, or merger.

In the early years of the industry, all carriers were subsidized; it was not difficult to increase the amount of support given to an ailing carrier to ensure the continuation of service. Subsidy was removed from trunks in the 1960s, though it continued into the next decade for the regionals. Low-cost loans required congressional action and were not likely occurrences. Route awards were the devices most often used to help ailing carriers, but were rarely successful. The committed policy of the CAB was to provide at least two competitors on a route.

When a third carrier was awarded the route to ease its financial problems, it resulted in all three carriers losing money. This is both reasonable and expected, for traffic demand tends to be inelastic, inching upward in small annual increments. New market stimulants in giveaways and pricing rarely increase the total market demand. Such an award would result in not two airlines sharing a market but now three lines vying for the same number of passengers.

The New York–Miami route is a case in point. It was adequately served by Eastern Air Lines (EAL) and National Airlines (NAL), achieving acceptable levels of profitability. When Northeast Airlines (NEA), a New England regional carrier, encountered extreme financial difficulties, the CAB awarded the New York–Miami route to Northeast.

Within a short period of time, all three airlines experienced marginal load factors and were unprofitable on the route. The route award did not solve Northeast's problems because of the added burden of capital requirements for long-range aircraft and its inability to gain a reasonable market share. Northeast was finally acquired by Delta Air Lines (DAL).

Within three months, the New England network was profitable although the New York–Miami run continued to lose money. Curiously enough, whenever one of the three carriers was removed from the market (by strike), the other two became immediately profitable.

Deregulation struck in 1978 and the capital-intensive airline industry became an arena of psychedelic experimentation without the necessary resources. Consider that more than 100 new airlines lifted off within five years after 1978 with only a modest incremental increase in total passenger demand.

Deregulation gave birth to a pack of aggressive competitors with fresh capital, non-union employees, and used aircraft. The mature, high-seniority, unionized airlines began to suffer losses when they met the price reductions, without the capital base to sustain them. Eating away at their equity, airlines finally extracted modifications from the unions and restructured their debt but they continued to suffer enormous losses until mergers began to take place by necessity. It is likely that the United States will be reshaped to less than half a dozen carriers at each level of service: majors, nationals, or regionals. As to whether a merger strengthens or weakens a carrier, is a subject for considerable debate.

In a macro-view of the merger ethic, it is clear that mergers have similar causative factors and outcomes, regardless of the industry. The characteristics of the merger process seem to transcend industries, and there is little reason to believe that this might not be consistent within the air transportation field.

A merger and acquisition study was completed by a group of European researchers recently, studying 765 merger cases spanning a 10-year period. They concluded that they could not find even a reasonable correlation between mergers and improved profit performance. Nor did their research show any increase in economies of scale or efficiencies that could be translated into lower prices. Granted that the firms studied were in Europe, nonetheless, the patterns were so consistent across a wide range of industries that it is apparent that the air transportation merger results would be similar. Over the last 30 years, the merger experience of U.S. carriers has not been satisfactory. For example, many years ago, Colonial Airlines, a New England regional carrier, merged with Eastern Airlines. Colonial was a borderline operator and there was considerable doubt as to its ability to survive; so EAL was approved as a merger partner. The alternative was to allow Colonial to fail; EAL would then have taken up the profitable routes and dropped the unprofitable ones. It would have resulted in layoffs and reduced service, unacceptable to the CAB. Eastern was permitted to acquire Colonial and required to take the routes, personnel, and equipment. Merging the unions proved to be bitter and expensive and the reorganization was costly.

Hindsight casts grave doubts as to whether the merger benefitted either carrier. When American Airlines (AAL) acquired American Export Airlines (AEA), which possessed overseas routes, the intention was to increase AAL's internal network allowing it to become an international airline. Within a few short years, AAL sold the international routes and equipment to Pan Am. Many of the routes were in parallel with Pan Am's. American's $6 million purchase was turned into an $18 million sale to Pan Am, after having suffered substantial losses. Pan Am eliminated a competitor for $18 million, a decision many airline analysts felt was unnecessary.

United Air Lines (UAL) acquired Capitol Air Lines, an Eastern carrier based in the nation's capital, from Slim Carmichael. Bitterness and recriminations still prevail within UAL, at both management and union levels with regards to that acquisition.

Pan Am's takeover of National Airlines in the late 1970s was considered synergistic because it held promise of better service for travelers. The two airlines served many of the same gateway airports, but Pan Am had almost no domestic routes and National had few overseas ones. National's stock was on the market in the low $20 range. Battling Texas International and Eastern, the price rose to about $50 a share, reaching an acquisition cost of $400 million cash.

In the midst of the takeover fight, deregulation occurred, and the irony is that Pan Am could have inexpensively obtained all of the domestic routes it wanted anyway. It was thought that the aircraft equipment and assets of National would equal the purchase price but the market dropped and dried up when it became glutted with aircraft. Then, instead of keeping the international carrier and trunk airline as profit centers, they were merged into one behemoth. The low-cost operation of the domestic operation immediately rose toward the high cost of the international operations and, in 1983, Pan Am lost about $50 million and was compelled to sell most of its assets.

Texas International earned about $60 million in a coup. It had bought National in the low $20 range and sold the stock to Pan Am. With that money cache, Frank Lorenzo acquired Continental Air Lines, which proved to be much more difficult to absorb than anticipated.

It has been said that to manage an airline is to manage its future. The management of an airline is the management of information. Airlines have grown so rapidly and the environment in which they operate is so dynamic that it is difficult to say that the organization of one airline at any particular time is the one in effect even a few months later. Organization, or the lack of it, has often been blamed for deficiencies of airline management. Moreover, the problems of an airline are unique and more complex than most businesses because of the 24-hours-a-day, 7-days-a-week, multiple-time-zones work schedule, and widely dispersed nature of their operation. To meet these challenges, an airline must be well organized and well managed. During the era of regulation and subsidy, airline labor relations were distorted because of government cutbacks in subsidy. With deregulation and new entrants, givebacks of contractual gains were widespread. Labor began to obtain ownership positions in carriers as trade-offs for contractual relief and productivity changes.

Mergers took place with greater frequency during deregulation. Most were not successful. *Fortune* magazine surveyed the almost $400 billion spent in the last 10 years in which some 23,000 companies were swallowed up. Then, almost 40 merger and acquisition specialists and security analysts were queried. Five out of seven mergers did not live up to expectations, and the Pan Am takeover of National Airlines in early 1980 was pegged as one of the seven biggest merger disasters in the last decade.

Among the lessons to be learned from the carnivorous behavior of airlines and corporations as they devour each other is that growth can better be achieved through internal expansion. History supports the thesis that mergers and acquisitions have not benefitted shareholders or employees, have not developed synergism, have not increased profitability, or served the public need and convenience. (See Figure 4.3 for principal airline mergers since 1950.)

Figure 4.3 Principal Mergers Since 1950

Year	Carrier(s) Merging	Surviving Carrier	Type of Service
1950	Arizona Airways Challenger	Monarch Airlines	Local service
	Monarch Airlines	Frontier Airlines	Local service
	American Overseas	Pan American	International
1952	Inland Airlines	Western Airlines	Trunklines
	Empire Airlines	West Coast Airlines	Local service
	Mid-Continent Airlines	Braniff Airways	Trunklines
1953	Chicago G Southern	Delta Airlines	Trunklines
1955	Pioneer Airlines	Continental Airlines	Local-trunk
1956	Colonial Airlines	Eastern Airlines	Trunklines
1961	Capitol Airlines	United Airlines	Trunklines
1967	Mackey Airlines	Eastern Airlines	Terr.-trunk
	Panagra	Braniff Airways	Trunk-intl
	Pacific Northern	Western Airlines	Alaskan-trunk
	Central Airlines	Frontier Airlines	Local service
1968	Wien Alaska	Northern Consolid.	Alaskan
	Bonanza	Air West	Local service
	Pacific		
	West Coast		
	Slick	Airlift Internat'l	Cargo
	Lake Central	Allegheny Airlines	Local service
1972	Mohawk	Allegheny Airlines	Local service
	Northeast	Delta Airlines	Trunklines
1980	National	Pan American	Major
	Southern	North Central	
1981	Hughes Airwest	Republic	Major
1982	Continental	Texas Air	Major-local
	Seaboard World	Flying Tiger	Cargo
1985	Continental	Texas Air	Major
	Eastern	Continental	
	People Express	Eastern	
1986	American	American	Major
	Air Cal		
	Delta	Delta	Major
	Western		
1987	USAir	USAir	Major
	Pacific Southwest		
	Piedmont		

Chapter 5

Marketing

Courtesy of Boeing Commercial Airplanes; Boeing 767-300

DESCRIPTION

Airline marketing is the process of matching the demands of present, potential, and future passengers with the supply offerings of an air carrier. The maximization of airline resources in markets entails the generating and controlling of marketing decisions as well as effective implementation.

Marketing, by and large, includes all of the associated activities related to the demand side of the coin. It is probabilistic and subject to high levels of volatility and uncertainty. Operations and costs, on the supply side, are deterministic and can be quantified with reasonable accuracy.

A market pair (origin and destination, or O & D) may be thought of as the dynamic arena between two cities. Each market pair may be specified as elastic or inelastic, as price sensitive or price insensitive, as growing or flat, and as being served by excess capacity or needing capacity.

During the first years after World War II, filling airline seats was a matter of selling. Passengers abounded and growth was robust. The view was commonly held that travelers would not "buy" an airline ticket unless the airline made a positive effort to "sell" it to them.

With the introduction of jets in the 1960s, supply (available seats) approached and then exceeded demand (revenue passengers). With the shift to excess capacity, selling was not enough. One marketing expert described the dichotomy thus: "Selling focuses on the needs of the seller; marketing on the demands of the buyer." The narrow "selling" notion proved unsatisfactory and was replaced with the broad "marketing" ethic. Selling and marketing are visually depicted in Figure 5.1.

Marketing is a conversion process: converting demand into revenue; converting available seats into revenue seats. There are some fundamental characteristics of the marketing process: (1) it is primarily the demand portion of the airline equation; (2) it is the revenue-producing side of the air carrier operating statement; (3) it interacts horizontally across every department in airlines; and (4) it cuts vertically through all layers of the company. The entire marketing process is shown in Figure 5.2.

Functionally, marketing is demand analysis, research, development, control, and planning. Marketing is determining the needs of passengers, the selling of tickets, passenger handling, and responding to feedback. It is reliability, scheduling, frequency, equipment, loyalty, attitude, and quality of service.

The marketing objective is to create "value" for a potential ticket buyer, and the goal should be achieved at a cost-benefit equilibrium. It

EMPHASIS   MECHANISM   OBJECTIVES

SELLING

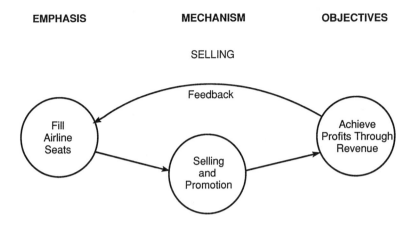

Figure 5.1 Selling and Marketing

MARKETING

Figure 5.2 Airline Marketing Process

must stop short of "marketing mania," which Theodore Levitt of Harvard refers to as being obsessively responsive to every fleeting whim of the buyer.

In the 1970s, four changes took place that severely affected the marketing of air transportation: (1) the introduction of wide-bodied aircraft; (2) rising costs of fuel and labor; (3) stagflation; and (4) deregulation.

Wide-bodied aircraft (two aisles) increased available seats well beyond the market's ability to fill them. Many routes suddenly had double capacity and more, with no increase in demand. The importance of the marketing function was highlighted as carriers fought to fill the empty seats.

The 1973 OPEC price hike caused an unexpected rise in fuel prices. The total cost per journey soared, and marketing groups were faced with the problem of higher load factors necessary to bring a flight to the break-even point.

What started as inflation after the Vietnam War developed into stagflation in the mid-1970s. Economic growth slowed, then came to an abrupt halt, and unemployment reached double digits. The public decreased its luxury spending; as a result, economy fares were introduced to try to maintain passenger load factors.

Under regulation in the United States a false competitive environment existed. Market options were tightly controlled by the CAB "in the public interest" and marketing decisions were distorted owing to government subsidy and political imperatives. It was difficult for a marketer to be innovative or inventive, for governmental interference often penalized successful operations. On the New York to Miami run, for example, both National and Eastern registered profits. When Northeast Airlines floundered on the brink of bankruptcy, the CAB awarded Northeast the New York-Miami authority as a third carrier on the hotly contested route. As a result, all three carriers began to lose money. (It is interesting to note that when one of the three went out on strike, the other two made money.)

Fare changes often took more than a year to obtain CAB approval and, if the fare request was an increase, it might take longer still. All competitive decisions requiring CAB approval became a matter of public record. Government regulation seriously hampered creative marketing.

Deregulation brought a revolutionary change to the practice of marketing. Tightly hamstrung for years by regulation, carriers were suddenly released to pursue marketing in its most elusive form. Airline marketers surged into an intensified period of experimentation and innovation. Few of the canons of regulated marketing applied. It was as Jonathan Livingston Seagull observed:

> We're free to go where we wish and to be what we are.

More than 100 new carriers jumped into markets in the first few years, armed with low wage structures, high productivity, non-union labor, and obsolescent aircraft leased at bargain-basement prices. They reintroduced the notion of "selling a seat."

Given the problems of increased capacity, high fuel costs, restrictive labor contracts, capital cost of wide-bodied equipment, and irrational behavior in the deregulated environment, survival and profitability depended upon an airline's ability to continuously identify, create, and exploit value in target markets. Low-cost predators were always in lockstep, snipping away at market share.

Airlines embarked upon highly erratic marketing innovations such as suicidal fare wars, giveaway inventions, route modifications, and a wide variety of untried scheduling experiments.

In 1981, more than 71% of all tickets sold were at an average of 46% discount, 78% in 1982 (46% discount), 82% in 1983 (48% discount), and 85% in 1988 (48% discount). Delta embarked upon a strategy to match all competitive fares. In an analysis of revenue in the summer of 1984, it was estimated that only 8% of Delta's fares were not discounted. One major carrier had 49 different fares on its transcontinental routes. Ron Wagner, American Express director, noted, "In 1979 there were 58,000 domestic airfares; today there are 4 million domestic fares, and between 40,000 and 60,000 change each day."

As air carriers sought to adapt themselves to the deregulated environment, *marketing management* became the magic term. What was needed were new ways of finding out, in advance, what the traveler wanted, could buy, and would buy. Moreover, new methods were needed to make the product available inexpensively. Marketing became the concern of all: the executive hierarchy, middle managers, and all employees.

It has been said that there is as much intuitive input to marketing as there is analytic input. As an example, the B-747 wide-bodied jet was produced at the behest of an airline president who intuitively saw a market need for a 300- to 400-passenger jumbojet. There is no evidence that the decision was based on analytical data.

For the first five years, the B-747 jumbojet was anything but a success. The engines did not perform according to specifications, and new ramps, ground service equipment, and maintenance facilities had to be designed and built at tremendous capital costs. It was estimated that Pan Am alone spent $18 million in the initial year's start-up, fine-tuning the B-747 to the airline's system. To make matters worse, passenger loads were inadequate. Airlines operating the jumbojet were strained financially until aircraft modifications, market growth, and learning experience made the B-747 a viable airliner.

Not long ago, Regent Air introduced an all first-class, super-luxury, transcontinental service on refurbished Boeing 727 trijets. Regent management perceived there was a market niche for a "Gucci" service. There proved not to be. At the other end of the scale, Hawaii Express management believed it had identified a market niche and introduced an all-economy, few-frills, low-cost 747 service from Los Angeles to Hawaii. When United and Western matched the fare, Hawaii Express was soon forced into Chapter 11 of the bankruptcy law.

Intuition and judgment played key roles in the market decisions. Demand was measured incorrectly and competitive response was grossly miscalculated. Such "marketing by gut feel" is not uncommon in air transportation. The lesson to be learned from these two stories is not to ignore hunches, but to test them with market research before attempting a costly implementation.

The industry continued to experience mega-jumps in capacity with a relatively flat growth in demand. The gap between the "haves" and the "have-nots" widened as strong carriers grew stronger, weak carriers watched their share of traffic erode, and low-cost predators chipped away in the most vulnerable markets.

The key to survival, noted Wall Street analyst Julius Maldutis, was in finding the right niche in the marketplace. He concluded that airlines which knew how, or learned how, to carve out a viable segment in the market were characterized by good management, a solid balance sheet, and the resources to purchase new equipment.

Abroad, government intervention and regulation was cumbersome, so bilateral negotiations stretched into years. In one case, Japan Air Lines negotiated 15 years with the United States for certain market rights.

Consider the international airline dilemma in present market decisions of future actions. Without warning, governments decide to freeze airline assets, after revenues have been earned. That prevents carriers from taking currency out of the country for tickets sold and freight billed within the country's borders.

For the IATA cartel of more than 100 members, the total blocked currency in 1983 was $790 million, approximately the equivalent of the net loss of its members. Each year the problem appears to get worse, and that is yet another uncertainty in marketing environment abroad. Even after choosing a viable market segment and taking the proper marketing course, it may all be stymied by a later blocked currency maneuver.

Finally, consider the dangers of varying exchange rates, which can wipe out profit from the time a niche is identified, market implemented, tickets sold, and processed for balance-of-payment funds.

THE MARKETING ORGANIZATION

The oldest and still most common basic organizational marketing format has at least three managers reporting to the vice-president–marketing. There are a number of permutations of the simple functional organization, with tasks being departmentalized, depending upon the size of the carrier and management's beliefs. The accompanying chart depicts the basic form:

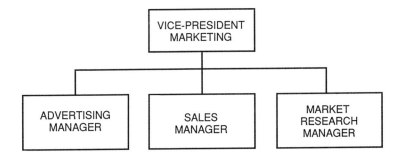

Advertising is such an important factor in the overall marketing picture that it is typically a separate department. In addition, it also covers promotion and publicity. As a general estimate, approximately 2% to 5% of an airline's total revenue is spent on advertising, promotion, and publicity.

Most carriers have *sales departments* that consist of three branches: outside sales (travel agents and commercial accounts); city sales offices; and ticket counters at the airport.

Market research stretches from analysis of traveler taste to organization strategy, and it employs disciplines ranging from psychology to advanced mathematics. A second area of research is market testing. A third is evaluation of feedback.

Marketing management is an adaptive logic. Airline marketing management must also be able to monitor and measure market positions, objectives, growth horizons, strategies, and long-range plans on a continuing basis. Marketing managers must be sensitive to feedback and the competitive environment, "not fixed in cement," and be capable of responding quickly. The selection of a management style has intrigued marketers for a number of years. Many marketing managers have tried to establish *a priori* rules for management to follow, but the best system is the one that works, for it is almost impossible to designate linear step-by-step rules of the road.

Stephen Quinto, founding president of Northeastern International Airways, observed, "The market manager must be a many splendored person; clairvoyant, researcher, creator, innovator, psychologist, psychiatrist, sociologist, economist, trend analyst, communicator, attorney at large, motivator, negotiator, visionary, crap shooter, and damned lucky." Rod Brandt, founder of Air Atlanta, advised, "If it works, don't fix it."

MARKETS

Markets are both nonspecific (or ambient) and specific. The nonspecific markets may require substantial capital investment to develop. Ambient markets are not specific to any one airline. A specific market, on the other hand, is a division into distinct and meaningful sets of travelers who might justify separate and distinct marketing ethics and services.

It might also be called market segmentation, the act of carving out a market niche through some differentiation. Studies have revealed a tendency of carriers to move from a specific market to an ambient market. People Express moved from a specific market as a regional carrier to an ambient market, national and international, as did Air Florida and Braniff.

In all three cases, specific marketing proved successful; the move to overall markets brought failure. One of the most foreboding challenges facing profitable specific market carriers is to grow while remaining in that niche.

Establishing a market follows at least seven steps:

1. *Evaluate* resources, goals, and objectives
2. *Identify* market demand and market-segment opportunities
3. *Match resources* to selected market niche
4. *Establish* a marketing organization and strategy
5. *Select* intervention and penetration level
6. *Develop* a long-range marketing plan
7. *Implement*, monitor *feedback,* and *respond*

Having pinpointed a target market segment by a selective process, marketing management then sets basic guidelines for penetration level, share capture, and implementation. A constant feedback and kneading must take place. Feedback is so important it has often been referred to as "The Breakfast of Champions."

Market Share

Airline market share is the percentage of a market held by a carrier in relationship to the total market. It can also be measured in segments (served market), or relatively (top three airlines), or comparatively (to another competitor).

In many industries it has been shown that Market Share (MS) of a total market is related to the Market Effort Share (MES). Market Effort

(ME) is the effort in money, time, variables, etc., that an airline puts into marketing its seats. This fundamental theorem of market share determination may be expressed as:

$$\frac{\text{Market Share}}{\text{(MS)}} = \frac{\text{Market Effort Share (MES)}}{\text{Total Market Effort (ME)}}$$

If two airlines have levels of service that are about equal, but spend different amounts of money on marketing the segment—$900,000 vs. $600,000—the first carrier would have a 66% Market Share:

$$\frac{\text{Market Share}}{\text{(MS)}} = \frac{\$900,000}{\$900,000 + \$600,000} = 66\%$$

Market Effort Share strategies often follow one of four directions:

1. Set a pricing strategy that optimizes market penetration even though it may dilute profits in the short run. (Most market experts hold that long-term profitability increases with Market Share. They also believe that markets are price sensitive, and low prices can stimulate growth. In addition, the cost per ASK (available seat kilometer) should be reduced with utilization of aircraft and route experience. Also, nonprofitable pricing will discourage competitive response.)
2. Increase MS by introducing new equipment, adding flight frequencies, more advertising, low-profit pricing, or upgrading service.
3. Maintain MS at existing level.
4. Undertake market skimming (value-added pricing). Use a skimming strategy to gain short-term profits by raising prices or decreasing advertising and service expenses and allowing MS to decline.

When market forces are in a relative steady state, the MS becomes a linear extension of the MES. In a regulated environment, for example, this would be consistent. In today's deregulated market, it is subject to erratic distortions, however. The interaction of supply and demand exhibits nonlinear fluctuations; has seasonal, daily, and hourly peaks; and is differentiated by unpredictable competitive tactics.

In most cases, carriers attaining a high share of the market are more profitable than their smaller (MS) competitive rivals. Airlines maintaining a dominant MS position tend to be highly profitable. That is supported by the history of two very profitable carriers, Northwest and Delta. They focus on oligopolistic markets which they can dominate. Some of the reasons are:

1. Economies of scale are a cost-effective method of achieving a higher rate of return.

2. A large MS allows improvements on the experience curve. (It has been found that total operating costs tend to decline by a constant percentage with each doubling of the cumulative output.)

3. Market Share dominance permits carriers to control pricing more effectively.

4. It is more economical to maintain a large MS than to try to improve one's position. (An increased MS capture can be extremely expensive and short-lived.)

Figure 5.3 graphically shows that, if one carrier increases its MES, assuming no competitive response, its MS will improve more than proportionally. There are qualifications, however. If the MES has been accomplished by reducing fares, total revenues may not increase. If MES has a cost impact, MS may improve but profits will likely drop. Should competition match the MES, shares will remain static. If the MES has been in increased capacity (available seat kilometers), load factors will drop and so will profits. If the carrier can reallocate its MES effectively with no additional capital expenditure, MS will improve, and so will profits.

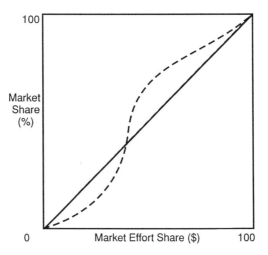

Figure 5.3   Market Share in Relation to Market Effort Share

In a comparison of markets of different sizes, it has been shown that the MS concept is more important in smaller markets than in larger ones. The Boston Consulting Group study clearly argues that it is more desirable to have a large MS in a small market than a small MS in a large market.

One of the explanations given is that, since there is less discretionary traveling on small (low density) routes, the market is more controllable. Also, fares may be at a higher yield in small markets because of diminished competition. Marketers have also discovered that MS emerges as more significant when the market is fragmented than when it is concentrated. The reason for this appears to be the diffusion of competition.

Having said all of this, it should be apparent how imprecise is the behavior of Market Share theory. Little has upset the marketers or done more to confuse the understanding of the theory than the inability to predict or control market forces. Finally, it is important to understand that MS analysis is valuable only when operating conditions are similar across carriers and, when this occurs, it is an extremely useful measure of market performance.

Positioning

Determining positioning in a market is one of the most crucial tasks facing marketing management. This involves deciding if the competitive niche is the best fit for the airline, assessing equipment/demand match, evaluating competition, predicting competitive response, and determining available resources.

There is a concept that relates to basic market position strategy put forward by the Boston Consulting Group (BCG). It is well worth discussion and proposes that marketing management has three choices: (1) growth; (2) preserving market share; and (3) liquidation. The critical variables are Market Position, Business Growth Rate, and Cost Experience. Adapting the BCG concept to air transportation, it is suggested that any airline can be positioned into one of four quadrants, as shown in Figure 5.4.

On the horizontal axis, from left to right, is the Marketing Effort Share as compared to the primary competitor. Positioning choices range from High to Parity to Low in terms of growth or buy-share strategies. On the vertical axis, from bottom (low) to top (high) is the industry growth rate.

The lower left square of the matrix shown in Figure 5.4 is the happy hunting ground of the CASH COW. Though the growth rate is not robust, airlines in that quadrant dominate or "own" the market, as

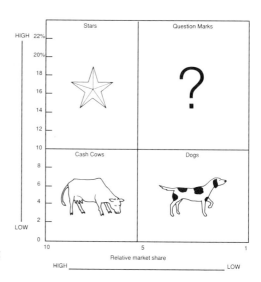

Figure 5.4 Boston Consulting
Group Business Portfolio Matrix

Delta and Northwest do in many cases. Large revenues are generated and it does not make economic sense to try to buy additional Market Share. It probably would be a good marketing strategy to protect MS and examine positioning in other markets.

In the upper right corner of the matrix is the QUESTION MARK, which possesses opposing characteristics. The market is small but growing such as Phoenix, Austin, and Atlantic City. The carrier positioned in this market has invested, carries a large debt, and is probably operating at a loss or barely breaking even. America West in Phoenix is such an example, as was AIA in Atlantic City. Airlines in this quadrant are also subject to predators, if the market grows. STARS is the designation in the upper left corner, a dominant position in a growing market such as American in Dallas, Southwest in Dallas/Love Field, and Midway Airlines in Chicago/Midway. In the lower right quadrant is the DOG, which is rightly named. It describes a poor position in a miserable market, British Airways to the Cameroons might so qualify. DOGS should be put to sleep, liquidated, and their resources funneled into better market positions.

In conclusion, the BCG marketing positions and concerted efforts should be to shift to the left side of the matrix in Figure 5.4. The DOGS should be abandoned rapidly, the QUESTION MARKS monitored carefully, and the bulk of the airline's resources moved to other segments.

Routes

The route decision is so interactive with Competitive Variables that it requires careful attention. By and large, oligopolistic routes are easier to market than are ones in which there is heavy competition. An often useful technique is to assign a market pair (origin & destination) as a profit center that establishes the route as a separate and measurable entity (with some limitations). The objective is to provide relatively unadulterated information as to the performance of the route.

If the route is international, negotiations must take place for landing rights. Frequency allotment, gate slots, and facility allocation can cause an excessive disadvantage. At Narita Airport in Tokyo, JAL parks in front of the terminal while many of the other international carriers are blocked miles away. After the U.S. and Philippine governments came to an agreement as to frequencies into Manila, not all American carriers were permitted frequency allotment into the foreign nation immediately. As in these cases, a route may be burdened with unbelievable Competitive Variable constraints.

Therefore, a route must be carefully weighed for today as well as tomorrow and, after it has been awarded and implemented, the route must be carefully monitored for competitive changes.

Pricing

At the root of the market equation is pricing. It is the element that creates revenue; its counterpart is cost, and the parallels are demand and supply.

There is general agreement that price determinants should reflect six major factors: demand, cost, competition, pricing objectives, profit, and management. Time after time, like lemmings going to sea, new carriers have entered major routes with low fares only to be matched within hours by major airlines. Markets tend to remain relatively inelastic, the total revenue decreases, and everyone loses. "The game," said one unenlightened chief executive, "is called 'stake out your turf.' The market strategy is to hold on to business by lowering fares where it will hurt a target rival and inflict minimal damage on oneself."

He never did explain how this occurs, and such specious reasoning has created havoc in yields for the industry. When operating scheduled service, People Express attempted predatory fare cutting on many routes. The majors matched the pricing within hours and it was a total failure as a market strategy. Then Capitol and World rolled the same dice. However, they leased old jets from Eastern and then competed on the same routes. When the aircraft were repossessed, Eastern had, in fact, subsidized its own competition.

The list goes on and on of carriers taking on heavily bankrolled majors in huge markets only to tumble. "We abhor fare wars," said one senior vice president–marketing at United, when announcing hefty fare cuts to match Continental's new fares in 1984. This is the announced response of the top majors: No one will undercut them. The sad fact is that carriers too often have made a shambles of pricing an airline seat.

A basic and simple pricing system is "one fare for one service," initiated most recently by Pacific Southwest Airlines. This is frequently the low-cost, all-coach, no-frills type of niche that prices a seat with little or no services. The second pricing system is more complicated. This is an aggregation of all of the random variables that affect the marketing of an airline service. A dynamic interaction takes place until a "value" is arrived at and a fare is determined, which is a measured compromise of all of the interacting variables in the marketplace plus a reasonable rate of return.

Thus, in this case, the airline fare is not an independent variable but rather an interdependent variable that is influenced by demand and supply factors. It is important to note that the larger the number of dependent variables, the more uncertain are the results.

Airlines have not yet settled on a balanced market pricing technique. A common mistake has been the excess use of marginal pricing of unused seats, which has impinged upon marginal utility. Air travelers play this game at least as well as the airlines, and it is easy for a traveling executive to grab a bargain-basement standby fare and hold his pricey positive space ticket for use on his next trip.

An understanding of marginal cost (MC) vs. average cost (AC) may shed some light on the machinations of discount pricing. Consider the formula

$$TOC = DOC + IOC$$

By definition, even when the airline produces zero output, it must honor its IOC (indirect operating costs) commitments. The IOC is the cost that is independent of output; therefore, it remains constant. Direct operating cost (DOC) is a variable cost and grows with each hour flown. Marginal cost is the increment of variable cost in the DOC to produce one seat. Average cost assigns equal units of total operating cost (TOC) to all seats sold (RPK).

As an example, if we assume that the ticket price is set to reach the break-even point at 65% load factor, then TOC = RPK (65%). In this case, the average cost would be equal units of TOC to 65% of the ASK (100%).

The two ways to measure the cost of an available seat are average cost plus marginal cost. Both can be calculated if the total input cost is known for various levels of seat output. Cost and volume are described in Figure 5.5 (using 65% load factor as an example).

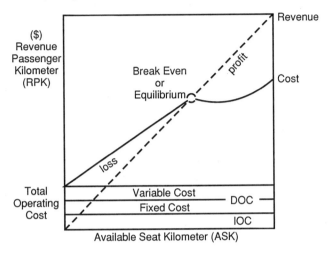

Figure 5.5   Average Cost vs. Marginal Cost

The difference between the revenue (broken line) and the cost (sold line) in Figure 5.5 is the loss or profit. The intersection of the two lines is the break-even point where TOC = RPK. Below this point, the solid line is the average cost of filling seats.

Though marginal pricing draws upon both the differential and contribution theorems, the pricing constraint is that the fare must cover marginal costs.

Contribution pricing is selecting a fare that will cover marginal costs plus as much of the TOC as possible, responsive to the elasticities of demand.

Differential pricing is designed to optimize benefits for a specific passenger group. This pricing technique is not widely used as a competitive option among airlines. Even though some competitive advantages do occur, differential pricing is directed toward various elasticities of demand. As a case in point, consider Possible Substitutionality (PS), another competitive option. When the (PS) is low, a high fare can be set. Differential pricing is often used when an oligopoly has market strength and the demand curve can be maximized by selective pricing.

Decision pricing in fare setting is established by management. Although economic determinants may be used to reach the final price, it is usually an arbitrary management decision. Even though pricing scenarios may be detailed and reasonable, with a ticket price clear and well documented, a different fare may be selected by management because of preferences, desires, expectations, or whims.

Clearly, the lowest risk option is to develop a set of fully allocated costs on a route segment. From there, the fare arrived at can be modified according to competitive factors, to reach revenues that represent a desired profit.

Empty seats cannot be warehoused and are lost revenue, which makes a worthwhile argument to consider how and at what price to fill unoccupied seats. The break-even fare and load factor cover the TOC. Therefore, additional passengers would have a low-cost impact. There would be negligible DOC (variable) in ticketing, meals, and service.

This makes a persuasive case for an airline to fill the empty seats at any price above the marginal cost, because of contribution. The kicker, which must be kept in mind, is that 3% to 7% of the full-priced passengers will switch to the marginal price, if offered.

Buyer Behavior

Demand in markets increases when the marketer is able to identify behavior and specific buyer preferences (not all of whom are receiving complete satisfaction from the current offerings or are unserved).

Academicians often break down the identifiable macro-aggregates of a market into the industry and consumer. There may be a slight dichotomy in understanding; the consumer is the end user in both cases as well as a subset of industry, but the consumer is not the buyer in both instances.

Travel agents and corporations comprise the industry buyers who purchase travel for business traveler consumers. These travel agent/corporate buyers represent as much as 75% of all airline tickets sold. In these cases, involvement and behavior are influenced by the buying situation and little or none of impulse. Industry buyers tend to be more schedule or convenience driven and less price sensitive. The industry buying process falls into five categories:

1. Need —   route, consumer's desired times, and
                         airport conveniences.

2. Search —   O&D, frequency, time of day, quality of
                     service, and price

3. Evaluation —   competitive options (including an air-
                     line's own schedule) and backup alter-
                     natives

4. Decision —   airline, flight, and service level

5. Perceptions —   good, bad, or indifferent

A consumer's purchase of an airline ticket can be either a routine buyer response behavior or the result of extensive evaluation and weighing of various airline offerings, one against the other. The consumer tends to be price sensitive, but also influenced by loyalty, advertising, and image.

To better understand the consumer, the marketer must ask:

1. Who is the traveler?
2. What are the traveler's needs?
3. Why does the traveler buy?
4. How does the traveler buy?
5. Who buys the ticket?
6. When does the buyer buy?
7. What is competition doing?

It is worth noting that a problem in small markets is that it is difficult to identify and exploit buyer behavioral characteristics. Many substructures of buyer behavior emerge in identifiable form with gains in market size. Although buyers may be broken into a number of subgroups, they are categorized more often according to geographic, demographic, psychographic, or behavioristic variables. To accomplish optimal segmentation, buyer behavior characteristics must be identifiable, available for data gathering, quantifiable, and qualifiable.

PLANNING

The purpose of a marketing plan is to convert available market niches into profitable outcomes. It must include available market options as well as the related costs, sales impact, and profit results. Finally, a market plan must be homogeneous, compatible and attainable within the scope of management's revenue/profit expectations and the firm's resources.

It is difficult to visualize a more active exercise than developing a market plan. The planner must take into account the airline's position in the market—as a dominator, challenger, trailer, or segmenter. In its position, the airline must decide whether to remain static or to maintain Market Share by growing at the market rate. It may elect to move to the next mode by increasing MS (dominator, the highest mode, simply expands the total market).

A market plan may attempt to expand the entire market by generating new travelers, new travel classes, or new uses; or it may simply increase travel in all categories. The market plan may shift resources into a variety of niches if management perceives this as a higher and better use of funds.

Market plans may be short range, lasting a year or less, or long range and experience a life cycle of about three to five years.

OPPORTUNITIES

Major market changes in relatively brief time frames can cause unpredictable opportunities to emerge. Identifying and exploiting marketing opportunities are vital marketing tasks.

The macro-environment of opportunities may be considered six ways:

1. Demography   (population shifts, income changes, birthrates, etc.)

2. Economy   (real income trends, inflation, savings/debt patterns)

3. Environment   (energy, pollution, fuel, noise, community constraints)

4. Technology   (aircraft design, delivery, flight operations, fleet mix, etc.)

5. Politics   (government controls, regulation, bilateral agreements, public interest)

6. Culture   (lifestyle, relationships, fulfillment, peer groups, language, habits, religion, beliefs, fashion)

The microsystem in the environment might be defined as:

1. Route demands (How large is the market?)
2. Competition (How much competition and how strong?)
3. Decision variables (How does this fit with our marketing plan and corporate objectives?)
4. Implementation (How long, how difficult, and how much?)
5. Buyer behavior (How will the market behave?)
6. Cost and revenue considerations (What are the cost factors and revenue forecasts?)

Marketing management is undergoing a radical transformation as it seeks to adapt itself to four monumental changes: the emergence and growth of more than 100 new airlines; highly competitive marketplaces; the impact of cost differentials (as much as 50% or more); and low-cost predatory invaders.

Increased competition, changes in personal, business, and pleasure travel habits, new route formations, new low-cost and efficient carriers, and the proliferation of pricing strategies are some of the factors compelling revisions in the methods of doing airline marketing. Nonunion jet carriers—utilizing innovative pricing schemes, with obsolescent but safe aircraft, and swarming over domestic and international markets to the delight of travelers—provide forceful reasons for deep rethinking of marketing.

Among the continuing problems of marketing management has been a reluctance to resist the temptation to add more capacity whenever it seems that traffic is picking up, then slashing fares when the seats remain empty. In 1983, it was estimated that 60% of the routes were priced at fares responsive to competition and almost completely unrelated to distance or cost! At today's tempo of change, no successful marketing policy can be successful unless there is a coupling of cost to pricing.

Regulation in the international environment continues to constrain both markets and creative marketing. About 80% of the carriers are government entities, have little control over their costs, and use pricing strategies that are reactive rather than generative.

Taking the theme a bit further, there is a new schism in marketing, one that calls on marketing management to see air travel in its totality instead of in fragments. Adjustment to today's reality is a task with which marketing management must cope if they wish their airline to survive. The marketplace will never be the same, nor will Bing Crosby ever come back for today it is Michael Jackson's time. The imaginative shuffling of costs among labor and airlines and traveler can lead to substantially greater profits. As Edward Barnet observed, "The mass market awaits those who adapt to it."

Chapter 6

Operations

Courtesy of British Aerospace PLC; British Aerospace 146-200

DESCRIPTION

Operations is charged with the responsibility of transporting passengers (and/or cargo) between points of origin and destination. The entire process covers three functions: the *intakes* are the revenue passengers arriving at ticket counters for check-in and boarding the aircraft (Ground Operations); the *conversion* takes them from the time they board the aircraft (Ground and Flight Operations), are carried to their destination (Flight Operations), and depart the aircraft (Flight and Ground Operations; and the *outputs* are when the passengers leave the airplane, and finally escape from the baggage claim area (Ground Operations). Maintenance is the support activity associated with keeping the equipment operating in all three functions of the conversion process.

Operations is on the supply side of the airline equation. Operations does not normally develop revenue (unless it sells its services).

At the apex of a present-day successful airline triangle is *profitability*. It is held up by marketing at one base angle and operations at the other. In the early years, operations was at the apex of the airline triangle and exhibited almost omnipotent power.

In the 1920s, airlines consisted of planes, aviators, pilotage, and mechanics. Carrying the mail was the mission of air transportation, for few passengers were boarded. Flying was relatively free of regulation and under the control of the pilots. To fly in command of a commercial airliner, an aviator needed only a pilot license, and training was a matter of a few takeoffs and bounces. Further, aviators became folk heroes reinforced by the feats of the airmail pioneers, barnstormers, crop dusters, and Charles Lindbergh, the "Lone Eagle."

It was difficult to question operational decisions as aviation was not well understood. It was held to be more an occult science replete with mysterious technology. The notable lack of federal regulation allowed a freedom of decision making, "according to the gospel" of each operations manager.

Because the emphasis rested on the technical aspects of flight, pilots prevailed in airline operations departments and rose to the leadership of most airlines. Howard Baker, Claire Chennault, Jack Frye, Eddie Rickenbacker, Bob Six, Mudhole Smith, Sig Wien, Lowell Yerex, and Woolman, Maytag, Quinto, Reeve, and Ward were a few of the aviators who led their airlines.

During the first two decades, airlines were driven by operations for four explicit reasons:

1. Mail was carried for most of the first decade;

2. Airlines were operational in nature;

3. Facilities and equipment were primitive;

4. There was virtually no regulation.

During aviation's second decade, the Roaring Thirties, the first modern airliner was introduced by the Boeing Commercial Aircraft Company. The B-247 increased the need for viable operating practices. As routes expanded and new aircraft were produced, such as the Douglas DC-3, operational issues developed that called for regulatory direction. The Civil Aeronautics Act of 1938 provided for complete federal control over every phase of interstate airline operation. Marketing and profitability continued to play vital roles.

The importance of how fleets of large aircraft were operated grew steadily and stemmed from the new technology and methods used during World War II as a result of the logistical demands of military support. In addition to long-range ferry flights, four-engine bombers were heavily loaded and flew missions that demanded operational optimization. Operations became more complicated and technological improvements increased dramatically.

After the war, the growth of operations surged in four distinct waves:

1. *First wave* in postwar 1940s and 1950s
   Surplus transport aircraft were readily available. Long-distance flying was introduced.

2. *Second wave* in the 1960s
   Jet Aircraft brought multiple operating problems.

3. *Third wave* in the 1970s
   Wide-bodied aircraft were introduced. Fuel prices escalated, the air-traffic controllers struck, and deregulation was instituted.

4. *Fourth wave* in the 1980s
   The full effect of deregulation was felt. By the middle of the decade, more than 200 new low-cost carriers of all categories were competing. The CAB was eliminated.

5. *Fifth wave* in the 1990s
   Business skills will be dominant. Pricing and yield management will be of paramount importance as carriers will succeed on their ability to derive information from data. Aircraft and market specialization will take place.

The first wave started after the war and was affected by the enormous amount of surplus equipment being made available for long-range flying. Airlines sprang up all over the world. Operating aircraft efficiently became a challenging task.

In the second wave, the jet tripled both the speed of airliners and the cost of operations; this forced the airlines to develop even more efficient operations techniques.

During the third wave, operations continued to dominate the thinking of airline management. Operations no longer simply referred to loading and flying aircraft, but grew to include anything and everything that could effect the cost of moving passengers.

Examples include flight dispatch and planning to determine optimal flight plans and economical cruising altitudes, and gate control and schedule coordination to make travel convenient for connecting passengers while economical for the airline.

In addition to their flying activities, large carriers began selling training services and maintenance capacity in order to provide additional income.

The full effect of deregulation was felt in the fourth wave, which focused on the supply side and operating skills. As the Jet Age progressed from the narrow-bodies to the wide-bodies, operational efficiencies became more prominent.

The industry came to grips with the insatiable trough in the guise of safety. Safety had to be equated in real terms of cost-benefits, and there rose a trade-off between reasonably safe operations and cost. The philosophy of "no matter what it costs, if it saves one life, it is worth it" was becoming unacceptable given the new era of cost competition and the impeccable safety record of carriers. One cold, analytical wag calculated the cost-benefit break-even value of a human life as $248,000 in 1984. This is such an arbitrary figure as to be debatable beyond general acceptance. It is intended to show that there may be a cutoff point not previously acknowledged. (After all, more people died in bathtubs in 1984 than in aircraft.)

It has been recognized that air travel is remarkably safe, and thus costly safety measures can go on and on virtually into infinity. This change in safety philosophy resulted in lower accepted levels of safety, schedule reliability, frequency, training, and maintenance, all of which had to be tolerated in the interest of costs and profitability. The payload of an airliner could be reduced to lower takeoff gross weights and thus lessen the danger of takeoffs, where most engine failures occur, but this would result in a penalty of lower revenues. Given the cost of a flight, it might require a substantial increase in fares.

As another example, an airline could record 100% on-time departures by assigning a standby aircraft on the ramp. The impact of such a backup aircraft might require a significant increase in fares. An equilibrium of about 95% on-time departures became an acceptable operational goal. The delicate balancing of delays vs. on-time departures, fully loaded takeoffs and potential engine failures, as well as a trade-off analysis in the matter of minimum operating weather for landing, was now recognized to have a cost that the public was not willing to bear. No longer were carriers on the government dole, which compensated them for operational inefficiencies in the "public" interest.

In the fourth wave, marketing became dominant and good management was the determinant of successful airlines. Deregulation was taking place in the United States while Europe clung to its regulation theory.

American carriers began to preoccupy themselves with cost-cutting measures, predatory pricing programs, and flashy marketing campaigns. Frank Lorenzo, of Texas Air, changed the character of the industry by turning it into a mass transit system.

In the fifth wave, which began in the 1990s, management rose to the fore. Airlines with first-rate managers continued to build up profits quarter after quarter. Poorly managed airlines went bankrupt, or were taken over. Marketing strategies varied widely, from low-cost, no-frills, to high-class business airlines. Specialization began to emerge, with airlines, aircraft, and on-board service matched to the markets being served.

When profitability and marketing reached their present status, operations had two more masters to serve. In addition to operational problems were service demands and profit objectives.

STRUCTURE

The termini of the airline conversion system lie at each end of *flight operations*. Flight operations has its own technology, which differentiates it from that which occurs on the ground. On each side of flight operations are *ground operations*, which consist of tasks of intake and outake. The third supporting function is *maintenance and engineering*, which is the nuts-and-bolts mechanism underlying the flight and ground activities.

The traditional operations management organization seems to follow the following pattern:

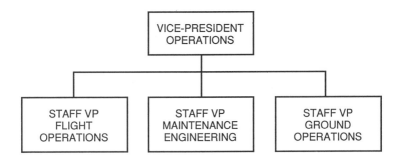

The basic operations structure centralizes control of systems operations, equipment routing, flight scheduling, crew scheduling, flight control, training, ground support, overhaul and maintenance, and operational public affairs.

There is a regular development of system service policies, or methods and procedures, to achieve standardization throughout the system. Charged with this responsibility is the Vice president operations, who strives to achieve five objectives:

1. Complete published schedules with highest reliability and performance at lowest cost;
2. Maintain highest possible safety performance at reasonable costs;
3. Maintain all ground and air equipment to meet flight schedules effectively, considering available resources;
4. Monitor performance to meet all government regulations;
5. Administer all departments cost-effectively.

Flight Operations

Under the department vice-president is the staff vice-president–flight operations. This person has the responsibility, authority, and accountability of flight personnel as well as flight operations policies, procedures, and standards. He (or she) ensures that the department complies with all Federal Aviation Regulations (FAR), operations specifications, and company policies. The staff vice-president is in charge of the flight deck crew members and directs their line flying operations. The VP monitors all aspects of route flying, evaluates schedule performance, and establishes guidelines that guarantee safety of flight, efficiency, and standardization. This individual provides leadership to flight crew members in accordance with the labor agreements, government regulations, and airline interests.

The vice-president—flight operations administers services, air/

ground communications, research, and related flight activities. He or she is involved in developing crew quotas, deciding when equipment must be purchased, and working with the Marketing Department to develop schedules.

Below the VP at the next tier are the chief pilot, chief flight dispatcher, and administrative assistant. This substructure varies with airlines owing to size, need, and other factors. Also included in flight operations might be flight service unless this function is placed within the province of the Marketing Department.

Maintenance and Engineering

The staff vice-president—maintenance and engineering is charged with the responsibility of maintaining flight and ground equipment in a safe and efficient condition; for coordinating the tasks within this department; for adherence to labor agreements; and for representing the airline as government liaison with respect to following regulations and changes in FAR relevant to maintenance and engineering.

The department often includes a major maintenance or overhaul facility for the repair, periodic checks, and overhaul of airframe, components, avionics, and engines. Within the departmental responsibilities are line-service activities and inspections, as well as research and engineering related to new flight equipment and upgrading the operating performance of the fleet. Inventory control in storage, classification, ordering, and issuance of company materials and supplies falls on this department. All vehicles, ground power, and support units are maintained by the department.

Ground-support equipment typically represents the equivalent value of about 20% of the cost of flight equipment. Commonality of fleet measurably reduces this percentage and cost.

At the next level are managers of maintenance, engineering, avionics, planning, research, supply and stores, facilities, overhaul, building and grounds, and properties operations.

Ground Operations

The staff vice-president—ground operations (SVP) is responsible for the administration of stations, ground facilities, and support equipment. The SVP also keeps close contact with various government officials (local, state, federal, and foreign) with respect to regulatory affairs. The SVP acts as the company's spokesperson in airport matters and is responsible for the airline's facilities at its stations. If an airport is a crew layover base, the SVP will negotiate all related contracts for hotels, meals, and transportation. The airport managers who report to the SVP are typically those in stations, ground service, technical, and airport facilities.

In sum, operations is responsible for production in the airline conversion process, from the time the passenger arrives at the check-in ticket counter at the airport, is carried to destination, and then leaves the baggage claim area.

Operations is also charged with operating and maintaining all of the equipment used to support the conversion, and it has always been a crucial element in the success of an airline. As stated previously, in aviation's early years, pilots and operations personnel dominated carriers. As aircraft and the industry became more sophisticated, government regulation evolved; marketing and professional management began to emerge as a vital force.

The problems of operations are complicated by the environment in which it functions. Air carriers are service organizations operating over land and sea, 24 hours a day, 7 days a week, 52 weeks a year, with widely dispersed time differentials.

When it is summer in the Northern Hemisphere, it is winter in the Southern Hemisphere. Weather systems in the Arctic are different from those in the tropics and must be understood for operational purposes. Ground services and equipment must be operated differently under the varying conditions in sleeting, subzero and sweltering, arid climates.

International carriers are buffeted by wars, revolutions, and civil strife that erupt unpredictably. Curfews and regulatory repression can be imposed without warning. Bangkok time is 12 hours different from San Francisco time and one day apart. It means that the normal working hours of the company's operations offices in Bangkok or Singapore or New Delhi may never be in sync with its headquarters. Holidays vary in these countries and are so numerous that important communications can be delayed or often impossible. Airways and airport facilities are not maintained with the same standards throughout the world. Thus, operations has an intricate and difficult task, particularly if the airline system is international.

Historically, the industry has gone through a complete metamorphosis about every 20 years, major changes every 10 years, and divertissements every 3 or 4 years. This has created problems for operations as its environment seems to be in constant flux.

In summary, *operations* can be broken down into three subsets:

Flight operations is the most critical, for its cost set includes flight crews, fuel, and equipment; perhaps 80% of the total operating cost.

Ground operations is the ground support for the airline and includes almost all of the activities that take place in the intake and output of the conversion process. The primary support system is *maintenance and engineering* (M&E), which keeps the fleet operating, as well as all of the ground equipment and overhaul facilities.

Maintenance and engineering has the nuts-and-bolts task of keeping all of the equipment operating. Delays can be costly; consequently, this department is extremely important in the airline scenario.

Operations is a highly technical responsibility and support activity. It must deliver the "product" to the passenger and is at least as important in air transportation as any other function.

Without marketing, there would be no revenue and no passengers.

Without operations, there would be no "product" to sell.

Chapter 7

Strategic Planning

Courtesy of Boeing Commercial Airplanes; Boeing 737-300

the allocation of resources, over time, that will
on of the airline and maximize profitability. The
cludes definition of the ends to be achieved, a time
nes to measure progress, and a description of how
t those milestones.

Strategic planning is systematic and should be formalized. It is optimal when written. It must take into account the power, political, and social forces within the airline.

Planning strategically is broad, conceptual, and consists of policy-making. It establishes corporate philosophy, expectations, and goals. The strategic plan (three to five years) includes a tactical plan for the short term (one year). It is objective oriented and an operational function. Also, a number of subset plans are aggregated to provide data for tactical and strategic plans. This description is clearly an oversimplification, but serves as an outline.

Strategic planning is made up of several interrelated steps, as seen in the following diagram.

Altogether, a coherent set of appraisal, mission, planning, and execution functions defines the scope of a strategic plan.

An appraisal and a mission statement are both essential before launching into the strategic planning process. Planning and execution are application oriented.

Strategic planning is perceived differently by various airlines. The most pragmatic strategic plan is often skewed. It has not changed and will not. Biases abound and plans are impacted by managers' personal beliefs, corporate culture, and intuitive convictions.

The rational structure recognizes all of the forces in the environment today and yesterday. The irrational factors are paradigms of management's aberrations of tomorrow.

Although a strategic plan may be intended to be an affirmation of reasonable goals, the higher planners are on the hierarchical ladder the more influential the bias. The strategic plan must also contain contingency alternatives, just in case.

Strategic planning, in a descriptive sense, does not call for the explicit definition of key words. In some cases, words appear to be in conflict, for, as S.I. Hayakawa observed, words are the sum of our experiences with them.

Given a potpourri of experiences, words strike different chords of understanding. Strategic planning could be bogged down by the insistence of absolute meanings because, unfortunately, there is not common usage in industry literature. It is important not to waste time in trivial pursuit.

For our purposes, *appraisal* is a measure of resources. *Mission* is management's concept of the airline's character, image, and what it wants to do.

Midway Airlines, for example, expressed its mission in a straightforward statement:

> Dedicated to providing quality air transportation at low fares; to increase the business volume, maximize efficiency, and minimize costs.

Goals are long-range future states that fulfill an airline's mission. A goal is more specific than a mission and can be stated in terms of markets, growth, and profitability.

Midway continued to cite its goals:

1. Low priced with two-tier unrestricted fares
2. Simplified service
   Single-class service
   No meals
   High-speed ticketing and reservation process
3. High productivity
   Employees
   Aircraft utilization
4. One aircraft type
   All DC-9 fleet
5. Hub-and-spoke route system
   Route structure constructed so that all flights operate to and from a hub airport, Midway in Chicago. This enables the company to provide one-stop service from spoke to spoke through the hub.
6. Peripheral airports

Midway Airport is uncongested and convenient to downtown Chicago. New York's La Guardia Field and Washington's National Airport are more accessible to the central business districts than other major airports in the respective metropolitan areas.

Objectives refer to short-term tasks to be achieved in the annual operating plan. *Execution* refers to the step-by-step implementation portion of the plan.

THE FOUR STEPS OF THE PLANNING PROCESS

Appraising the corporate identity is the first step in the planning process, and *appraisal* centers around the question "Who are we?"

It is axiomatic for airline management to identify strengths and weaknesses: (who they are, what business they are in, and what the available resources are) before determining what business they want to be in and how they are going to get there.

Specific resources are physical assets, equipment (both ground and flight), routes, landing rights, gate positions, people, skills, competitive advantages, barriers to entry to competition, marketing, operations, and finance.

Nonspecific resources are the corporate culture, changing goals, operating variances, competitive responses, hostile environments, operational limitations, government regulations, assets management, the unpredictable economy, and the irrational behavior of competitors in the future.

Developing the *mission* or purpose of the company is the second step in the planning process. The question to be answered here is "Where are we going?"

A mission statement underscores long-range goals. It is a holographic description of what the airline is all about. Foremost, a mission statement must target an achievable horizon, not the blue sky. Extolling a euphoric mission in a twilight zone filled with pipe dreams rather than hard-hat reality can lead to frustration, suboptimization, angst, and trigger a serious shortfall.

The intended outcome of the mission statement is mutual understanding and consensus. Similar to the Calvinist covenants that bound early New England communities tightly, mission statements nurture pride and loyalty and create a sense of oneness. Given these notions, it is not uncommon for the formulation of a mission statement to take longer than writing the strategic plan. Perhaps this is what Dwight D.

Eisenhower meant when he wrote: "Plans are nothing; ⌐ everything."

In a lighter vein, one of the all-time great mission ⌐ could be found on the wall of a cobbler's shop in venerate⌐ Square. The framed statement weathers the test of the passing years:

> We are dedicated to the saving of soles, heeling, and administering to the dying.

The actual *planning* is the third of the fourth steps in the strategic planning process. The question being answered at this stage is "How do we get there?"

Strategic planning is a road map with the direction brightly indicated. Questions of all kinds arise. In which arena can the carrier operate effectively: international, national, regional, or commuter? Strategic planning centers on questions and provides a consistent guide to be used to develop answers to these problems.

Many trade-offs must be considered by management in all areas of airline service, the most basic being:

1. Market share increase vs. new market development
2. Profits vs. market share
3. Short-term dividends vs. long-range profits
4. Related vs. unrelated opportunities
5. Profit goals vs. societal goals
6. High risk vs. low risk
7. Dynamism vs. stability
8. Debt vs. equity
9. Consensus vs. divisiveness
10. Quality vs. profit
11. Market leader vs. follower
12. Low profile vs. high profile

Management selects the marketplace in which the airline is perceived to be most robust; it positions the carrier and looks to the future. Management personnel evaluate market demands, operational factors, and financial constraints. They carefully define which segment to serve, the niche opportunity, and how to exploit the market. All in all, a well-constructed strategic plan must be a credible valuation of resources, a "do-able" strategy, and an optimal tactical course.

In developing the strategic plan, top-down, bottom-up, or goals-down/plans-up methods can be used. In a *top-down* planning environment the top management establishes goals and plans for all levels of the airline. This is an adaptation from the military. Senior officers prepared the plans and the troops carried them out. This style of planning requires no formal approval.

With the *bottom-up* approach, corporate divisions prepare subsets of the strategic plan. Segments are aggregated when completed to construct an overall plan that is then subject either to approval or rejection by top management.

The *goals-down/plans-up* approach is the most commonly used method of all. In this scheme the top management sets the tone of the strategy, explores opportunities, and defines objectives for the coming year. The departments and divisions of the airline react to these objectives by developing pieces of the strategic plan for their areas, and then these are aggregated at the top management level.

Emphasis is on the annual operating plan. The reason is because the airline industry is short-term oriented and cash-flow intensive. Desired outcomes happen because managers and top management establish a dialogue to proactively develop the strategic plan and tactical operating plan. The inputs into the planning process take into account past experiences, present conditions, the futurity of markets, available resources, competitive actions, economic indicators, and the uncertainty of environments.

The fourth step is the *execution*. It is the response to the "How to Get the Job Done" question. This step will be discussed further later in this chapter.

CONTINGENCY PLANNING

A viable strategic plan also contains contingency choices at each branch of the critical path. Knee-jerk reactions end up in inconsistent actions and suboptimal results. Fallback options are vital.

Managers examine every resource to develop a firm database and a set of reasonable assumptions. A noisy, symbiotic dialogue should take place among marketing, operations, and finance, overseen by top management.

Marketing describes the demand, defines a market niche, outlines strategy in type of service, highlights market characteristics, develops a capacity penetration plan, analyzes the competition, plans revenues, and maps a path to profitability.

Operations crunches numbers to determine whether or not it can support the plan. Operations is on the ground floor of every strategic planning decision. Some details left to operations include determining fleet strategy in number and type of aircraft; describing the potential for fleet expansion; deciding on engine types and spares requirements; and determining crew and training requirements to meet growth scheduled in the strategic plan.

Finance is the deep pockets and "the buck stops here" department. Each variable must have a cost estimate developed for it by the finance group. Equipment, fuel, maintenance, schedule, labor, routes, landing fees, ground services and the like—all have to have their costs carefully examined to be sure that the carrier can financially support the strategic plan. The airline might be highly leveraged and the opportunities to raise the necessary capital to embark on an ambitious competitive campaign might be quite limited.

ANNUAL OPERATING PLAN

A famous general observed that a long-range plan is a strategy for winning a war while a short-term plan describes the tactics needed to win a battle. Short-term planning centers on defining the current problems and how they can be solved. Tactical planning for the short term is selecting the means to accomplish objectives.

Change is rarely a major problem as the time span is selected to be short enough to avoid such problems. Typically, the short-term plan is an operational management responsibility.

Market Planning

Marketing is, far and away, the most important factor in strategic planning. Marketing options can be broken into three classifications: undifferentiated marketing, differentiated marketing, and market niching.

Undifferentiated marketing suggests developing a single broad marketing effort to appeal to all potential travelers. It is most consistent when markets are relatively stable, and equipment and scheduling are standardized.

For example, when routes are certain lengths, one type of aircraft will best serve that route pair. If there is heavy demand, the market probably supports a common price, frequency, and schedule. Los Angeles to Honolulu is served by wide-body jumbo jets. It is difficult to differentiate the carriers serving this market by either price or service.

Differentiated marketing occurs when a carrier operates in markets that are diverse and buyer behavior differs appreciably. In this case, market differentiation is crucial. Differentiated marketing means the marketing method is tailored to best fit the market structure. Transcontinental flights can offer price differentiation by flying nonstop, direct (same aircraft but with a stop), or connection (changing aircraft at the stop). There can be night flights, stopover rights enroute, and "sleeper" service. If a market responds to the differentiation, it can result in a robust competitive advantage for the innovative carrier.

Market niching is a strategy of entrenching a carrier inside small, uneconomic market segments and developing them into viable opportunities. If an airline develops a market niche and matches its structure to the demands of that segment, it can well become a stronghold for that carrier.

A prime example of this phenomenon is the success of Southwest Airlines in Texas commuter markets. Southwest developed these markets as low-cost commuter markets and has done quite well serving them with Boeing 737 aircraft. Customers have grown loyal, and other carriers cannot seem to take these markets away from Southwest. In spite of giants like Continental serving the same cities from Houston, and American and Delta competing out of Dallas, Southwest Airlines has built itself a comfortable market niche.

The strategic plan must carefully pick the market apart. Is it a large market? A niche? A new market? Does an opportunity exist for a large market share, or can a carrier be profitable by achieving a small one? Passenger demand must be quantified. The optimal schedules, equipment, and pricing are of little value if there is not a serviceable demand. It is also true that a supply of seats will not create passenger demand.

Marketing puts together a plan to analyze markets. The marketing short-term objectives look at markets to be served in the coming year; demand, schedules, frequencies, equipment, utilization, pricing, revenue generation, timing, competition, and advertising are all considered. By and large, marketing is oriented to optimize resources for marketing gains and profitability.

Sales Forecasting

The sales forecast is useful in establishing sales targets, identifying purchasing needs, specifying advertising and promotion budgets, evaluating price levels, and targeting market share. Sales forecasting can be either regressive or aggressive. The regressive method uses past data for sales linear projections, whereas aggressive forecasts rely on present information and future orders. These methods are not mutually exclusive, and some airlines use both.

Sales estimates can be gathered from field sales representatives. The sales staff is in the best position to estimate current passenger demand. They see competitive responses firsthand, and the results of airline strategies are most visable to the field sales staff.

Consensus forecasting calls for marketing managers and top management to discuss sales potentials until a consensus is reached. Opinions are expressed and closely analyzed until a rational forecast can be constructed. Final forecasts are cross-checked against customer surveys and any available industry data.

Forecasting errors demand careful attention. Though sales forecasting might use precise methods of measuring demand, it produces mere estimates of passenger behavior in the future. Thus, best-case/worst-case scenarios are often constructed to compensate for future deviations. Presented with these adjusted data, management must make a final sales forecast based on the estimates.

The final forecast must be free from moonshot enthusiasm or doom-and-gloom pessimism. In the end, top management relies on the information, then makes an educated judgment.

Forecasting efforts can also be performed for particular purposes in different phases of the planning process. All sales forecasts should be made in dollars earned or tickets sold.

Schedule Planning

Schedules are hydra-headed. On one side the field will demand as much capacity and frequency as possible on every route, so the various airport managers can provide the best possible schedule to their customers. Competition with other carriers also depends on nonstop service, many departures, and the latest and largest equipment.

On the other side, schedules are the slave of equipment availability, crew resources, fleet utilization constraints, and economic considerations. Schedule is a critical competitive tool that can make or break a carrier in a particular city.

Recognizing the importance of schedule dominance has led to "capacity wars" on many routes. Consider Chicago–Minneapolis in late 1987 and early 1988. Chicago is the major hub for United Airlines at O'Hare Airport, and for Midway Airlines at Midway Airport. At the other end of the route, Northwest Airlines has its major operation in Minneapolis. None of these carriers wanted to be "out-frequencied" at their own hub, so each began building up service on the profitable Chicago–Minneapolis route. United and Midway added flights out of their hubs, with Northwest adding them to protect Minneapolis.

Before long, each carrier was flying this route with hourly service, 16 to 18 flights daily. Combined with service by a few other airlines with Chicago operations, the number of flights between Chicago and Minneapolis grew to more than 70 a day. Needless to say, all carriers began losing money on this route.

Fleet Planning

Fleet planning recognizes loads, schedules, positioning, frequency, maintenance, and equipment growth. It is necessary to evaluate the fleet in terms of both present and anticipated markets, load factors, and route segments. Growth is a consideration as well in parts support, crew qualification, and maintenance. The number of hours that an aircraft flies each day (utilization) is extremely important. Utilization targets typically run about 10 hours daily at 60% or higher load factors. Operations estimates resources needed, assesses those available, and develops a plan to coordinate with the marketing department to serve target segments.

Utilization is a significant variable. For large airlines like American and United, every extra two minutes of daily average utilization can be equivalent to an additional aircraft. However, there is a trap in this logic. Utilization can be added beyond the break-even point.

Additional hours are only valuable if they generate economic loads, if they can be blended smoothly with existing schedules and fleet positioning, and if they will not cause equipment to be out of service at other times because of reduced overnight maintenance.

Pricing

A pivotal factor in the market equation is pricing each seat. Pricing takes place as markets are being defined, their potential assessed, seats made available, marketing channels selected, and promotional programs implemented. Pricing and total trip revenues are paramount considerations when new routes are planned, for revenues are the airline's life blood. In the strategic planning process, the price is measured as the average revenue per passenger, or total revenue divided by total tickets sold for that route.

Ticket prices must cover costs while at the same time be responsive to market demand. Failing at either of these missions might seriously damage an airline's cash flow and market share.

Pricing must anticipate competitive response before an action is taken. To price above equal competition is not possible, but premiums for nonstop service, for higher frequency, or for a better frequent traveler program might be supported.

Sunworld Airlines operated routes out of Las Vegas, Nevada, in direct competition with America West. These flights were operated with identical equipment, comparable service, and on similar schedules. The only difference was Sunworld's higher price. Ultimately, Sunworld withdrew from these markets and faded into oblivion.

Pricing impacts the degree of market penetration, level of competition, and type of service available between two cities. The low-cost carrier sets the prevailing fares at all price-sensitive levels including both leisure and business fares. The only fares that seem not to be price sensitive are the premium First Class and Business Class fares. Although competing carriers usually match these fares as well, they are generally as high as anyone can stand to set them at nearly 150% of full Coach.

Revenue Planning

Total trip revenue is the key to revenue planning. Ideally, total trip revenue is the average revenue per passenger multiplied by the number of seats sold. With all of the competitive pricing schemes available it is not possible to estimate the total trip revenue in advance. For example, in 1987, it was estimated that 90% of all airline tickets were sold at some type of discount. The consideration must be the total aircraft revenue per leg, and an estimate of this is merely a shot in the dark.

Financial Planning

The declared responsibility of financial management is to maximize shareholder equity. Its fundamental responsibility is to keep the airline from going broke. Finance monitors debts and assets, calculates future funding needs in light of proposed service offerings, and estimates revenue streams.

A budget is derived and is simply a tactical financial plan, a list of goals and a tool for measurement. It is useful for both planning and control, and it starts with a set of known performance standards. A good budget system will recognize that some factors lie outside of the airline's control. Therefore, flexible budgets must be set up for different departments assuming diverse levels of performance.

Maintenance Planning

Maintenance develops its own annual operating plan to forecast overhauls, downtime, and unscheduled maintenance. There must be a timetable for periodic maintenance, mechanical crew training and availability, contractual constraints, government regulations, and aircraft utilization.

EXECUTION: HOW TO GET THE JOB DONE

The bottom line in the strategic planning process is *execution* once the plan is complete. The plan-in-action is a vital guideline for measuring success of the process. During execution, the plans are carefully monitored and evaluated every 30 to 90 days. The budget figures are compared with actual results. This is the control phase of the budget system. It is a critical discipline in well-operated airlines. Some managers suggest that plan deviations should be acted upon whenever conditions vary by more than 5% in any of the critical areas.

An example of carefully planned execution can be seen in the success of the Green Bay Packers football team under the guidance of Coach Vince Lombardi. Green Bay dominated professional football for years, even though most players in the league possessed similar skills and were in the same physical condition. The plays the Packers used were not magical, nor did they have a fantastic defensive line. What made the Packers successful was careful execution of well-considered plans!

The truth is, strategic plans are only a set of instructions. Good execution can make a success of a poor plan, while poor execution will surely cause a fine plan to fail. A military academy in California lived by a motto that might well be applied to the strategic planning process:

Deeds, not words.

WHAT STRATEGIC PLANNING IS NOT

Strategic planning is not a shopping cart of scientific techniques. It is analytical thinking and a commitment of resources to action. Strategic planning is not the application of absolute formulae to business decisions. It is thought, analysis, bias, judgment, and often guesswork that lead the planners back to square one.

Planning is decidedly not to follow cookbook-style methods in the quest for accuracy. Sometimes, numerical solutions serve little or no purpose in the real decisions to be made. In isolated cases, quantification can provide validation or support of particular parameters or ideas, but important issues often defy the rigor of mathematics.

Strategic planning is not second-guessing coming events. The mightiest blow might be struck for the cause of strategic planning because we cannot predict what the future will bring. Strategic planning deals with a range of probable futures and the balance of risk with return in the most likely of them.

Wrestling with a rational system and many certainties is not strategic planning. Rather, without the umbrella of government regulation, management deals directly with irrationality and uncertainty. It is not precise, yet for discovery and enhancement of present resources, strategic planning is the only way.

Strategic planning does not substitute for judgment or science for managers. The process is an art form, far afield from the laboratory. Skills, courage, experience, intuition, and resourcefulness are not replaced but enhanced by the process.

Finally, strategic planning is not a way to eliminate risk or to minimize it. The plan simply allows management to select from the menu of risky courses of action one that best matches the goals and desires of the company.

SUMMARY

To recap, strategic planning is necessary. This is supported by the performance of most successful airlines including American, Delta, United, USAir, and others. They respect the planning function and view it as a means of enhancing profitability.

Specifically, the planning process must be undertaken to develop a supply/demand scenario to ensure the best use of the carrier's resources. It must offer a methodology for careful and continuing feedback, providing for rapid and responsive modifications.

Contingencies must be explored at every branch of the plan. The environment in which airlines exist is in constant flux. Thus the most carefully sculpted strategic plan will be modified as a result of unforeseen events. A strategic plan is never a hard-and-fast rule.

What is significant is that planning is, first of all, a description of objectives so that everyone involved understands the goals of the airline.

Second, it is a description of the systematic and purposeful task of attaining these goals.

Third, it ensures that each component of the plan starts at a zero reference point, sloughing off the biases of yesterday.

Fourth, it gives weight to the time dimension, an understanding that the decisions made by management today contain the possibility that the future will bring fragile changes.

Fifth, it is a recognition that feedback, evaluation, and modification of the airline's environment and state of being must be continually monitored so the plans can be revised and re-revised as situations change.

Finally, all is for nothing if the execution is less than first-rate.

Chapter 8

Equipment
Selection

Courtesy of Fokker Aircraft; Fokker 50

DESCRIPTION

Airline fleet planning and equipment selection involve determining present fleet resources and future equipment needs within the context of company objectives. Consideration of the appropriate equipment is one of air transport management's most difficult tasks, and the process is intended to render a decision that reflects an optimal balance among operations (equipment), marketing (demand), and finance (resources). See Figure 8.1.

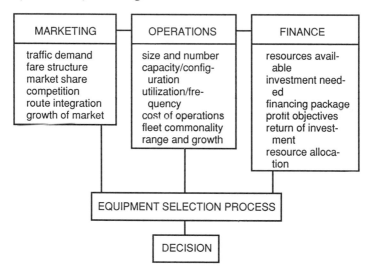

Figure 8.1 Major Equipment Selection Factors

The equipment selection process is based on substantial input information. Current fleet including engines, airframe, and ground support must be considered as well as aircraft performance, operating objectives, initial schedules, route structures, and competitive airlines' positions.

The composition of the market demand and how it fits in the present airline system and plans for future growth should be analyzed. Also, the financial resources of the firm and the financial conditions of the equipment acquisition will impact the ultimate decision.

The marketing group must identify the passengers' demands in terms of quantity and type of service (or quality) and then clearly describe the market strategy. Fleet planners must then translate this information into an equipment recommendation.

Charged with finding aircraft that best meet these criteria, fleet planners must consider the models available, fleet mix, capacity, configuration, growth potential of subsequent models, parts support, airport facilities, ground equipment, and a host of other considerations in making their decision.

Few nonspecialists appreciate the scope of selecting equipment for the addition to or modernization of a fleet. There are more than 100 different versions of wide-bodied jets, for instance, each with its own performance characteristics and economies of operation. They include about 65 alternatives of 10 Boeing 747 airframes with 10 types of engines.

The DC-10 and L-1011 aircraft could be tailored to any one of hundreds of combinations in engines, airframes, and interiors. The many options possible in newer-generation aircraft like the Airbus A-300, Boeing 757/767, and MD-11 are mind-boggling. Add to these hardware offerings the unpredictable nature of fuel costs (differing from airport to airport), increasing labor costs, training requirements, ground support systems, parts inventories, airways usage constraints, landing tariffs, gate slots, overhaul availability, and known management biases, one can quickly understand that the equipment selection procedure is extremely messy.

Furthermore, management's equipment selection decision may appear to ignore the facts as presented, yet the "mating dance" process must take place to provide management with an insightful analysis of equipment choices. A poorly weighted selection can result in the wrong aircraft for the right market and threaten the viability of the airline.

As an example, Hawaiian Airlines and Aloha Airlines, both serving Hawaiian interisland routes, took the stance that equipment differentiation was an imperative.

When Hawaiian bought the Convair 580 turboprop twin, Aloha countered with the Fairchild F-27 turboprop. Hawaiian upgraded to four-engine DC-6B's and Aloha bought the Vickers Viscount. When the Jet Age hit Hawaii, Hawaiian selected McDonnell Douglas DC-9 twinjets and Aloha opted for Boeing 737 twinjets. Economies of scale would have resulted had both regional carriers operated the same aircraft type in the Hawaiian Islands. Management's irrevocable competitive strategy, however, was a preoccupation with differentiating equipment. Both carriers were only marginally successful as a result.

STRUCTURE

It is worthwhile to comment on the many factors that interact, to give some sense of the complexity and trace the difficult path followed to equipment selection. T.S. Eliot said that poetry begins to communicate long before it is understood. That also describes the delicate communications, sensing, signals, and evaluating of the three major forces that challenge management to "understand" their point of view on equipment. Operations, marketing, and finance perform careful analyses and, after all is said and done, management retains the prerogative to make the equipment selection. Why management in the face of the evidence?

Clemenceau, the French statesman, once observed that the conduct of war was too serious to be entrusted to the generals. Thus the selection of equipment is too important, it is said by managers, to be left to the fleet planners. There is validity to that notion, for management holds the macro-view of company strategy that has broader implications than the fleet planning analyses. As a matter of fact, not one of the interacting forces of marketing, operations, or finance should dominate if a truly optimal result is desired.

Playing a pivotal role in this scenario is the *accuracy* of assumptions, for it is possible to support almost any decisions if they are biased. For example, the Marketing Department forecasts that a series of city pairs will generate 90 to 100 passengers per flight and grow to 130 to 150 passengers in two years. Based on that assumption the airline acquires 180-passenger aircraft instead of available 125-seat planes. However, the market produces fewer than 90 passengers because of unexpected competition; the demand remains somewhat flat, and growth is slow. The fixed costs of the excess capacity and increased direct operating costs of the larger aircraft can consume an airline's precious resources in a short period of time.

The decision was valid, but the assumptions were flawed. It might well be the "right" aircraft but it is the "wrong" market and the solution of the 180-seat aircraft was not robust enough to make it a viable choice in an uncertain market situation. Whereas the cost per available seat kilometer was lower for the larger aircraft, the total trip or cycle cost was higher. In this case, either aircraft (125-seat or 180-seat) would have served the prevailing market and would have allowed for some growth.

If the market expanded rapidly, it could be accommodated with additional small aircraft, less efficient than the larger aircraft but also far less risky. If the market did not grow and demand remained inelastic, the smaller aircraft would have been more effective.

This is a simplistic example of *overcapacity*, the malaise of the air transportation industry. Theorists who advance economies of scale suggest that the larger the aircraft the greater the savings per seat kilometer.

In practice it has not held true, for this assumes the same percentage load factors for all sizes of aircraft. Because of the cyclical nature of air travel, the risk is markedly reduced with the smaller aircraft. With smaller aircraft the schedule is more flexible because lower fixed costs allow flights to be cancelled when necessary or more schedules added at peak times.

This is exemplified by the phenomenon of the wide-bodied carriers having been less profitable in many markets than twinjet operators. This further highlights the matching of aircraft to markets (supply to demand), a critical issue requiring precision. It is obvious that the equipment selection process is sophisticated, highly interactive, and uncertain; subjective judgments can be disastrous. Given the variables described above, and in view of the fact that organizational structures, corporate objectives, management styles, and planning techniques differ from airline to airline, no single qualitative or quantitative equipment selection ethic can be defined. Each carrier develops its own best method to suit its own personality. Certain selection procedures do have *some* common characteristics, and these are illustrated in Figure 8.2.

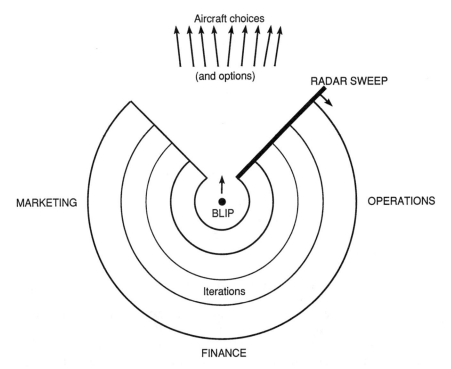

Figure 8.2 Equipment Selection Process

Figure 8.2 represents a typical equipment selection process. This process can be visualized as a blip on a radar screen, which starts from the center and moves outward. The radar sweep-hand continues to rotate and interrogate the various sectors as the decision blip migrates outward through each iteration toward the outer rim and the decision exit. As the sweep-hand encounters new information, it may cause the blip to alter course. As the decision blip moves toward the outer scale, it is repeatedly influenced and its course changed by new inputs from operations, finance, and marketing. By the time the decision blip reaches the exit decision gate, it should have been so fine-tuned that the cumulative output is a clear equipment choice.

The process represents the combined contributions of the following departments:

1. Planning and Scheduling   — Aircraft mission definition
   Capacity requirements
   Range and utilization
   Determination of fleet mix
2. Marketing   — Interior configuration
   Seating
   Baggage needs
   Galley requirements
   Entertainment systems
3. Operations   — Flight
   Flight characteristics
   Fleet commonality
   Cockpit layout
   Instrumentation
   Avionics
   Flight training
   Technical
   Aircraft/engine
   Spares inventory
   Maintenance training
4. Finance   — Evaluation
   Planning
   Contract negotiations

The three major factors just mentioned and their subsets all play key roles in the equipment selection process. If these are not subjected to an iterative process, the aircraft chosen may not be the one best suited for the market. In the following sections we consider each factor in detail.

DEMAND

Traffic demand is the pivotal element in equipment selection. Aircraft are acquired to meet passenger demand in a *facile princeps* manner, to provide the maximum revenue at the lowest cost.

Providing an optimal balance of available *seats* to *demand* is the final objective of equipment selection. Therefore, defining the traffic demand is the first step in the equipment selection procedure. The interdependence of various factors and their relative values, must be well understood. Typically, the following steps are involved:

1. Marketing defines the target market pairs and segment lengths; determines the entire route structure, estimates *traffic demand*, competitive factors, and broadly quantifies service needs. Estimates are based on historical information, competitors' schedules, "ramp counts" of competitors' traffic, and, in some cases, data purchased from aviation consulting and research firms.
2. The *fare structure* is determined by existing fares, competition, TOC, and penetration strategy.
3. *Market share goals* come from forecasts, schedules, route structure, and traffic performance simulation with a market share capture evolution.
4. Competition is identified as present airlines in the market and anticipated competitive reaction.
5. Route integration is the allocation of aircraft to itineraries in order to maximize total system profit. It is a "fit" analysis.
6. Growth of market is a linear extension of historical data, elasticity of demand, and macro-view of the future.

Traffic behavior is sensitive to irrational variables, and seems almost capricious at times. It is subject to directional imbalances, peaks and valleys in times of day, weeks, months, and seasons, and the unforeseeable whims of competitors. As a result, a flight might have a 95% load factor in one direction with a 10% load factor in the other. During peak times, load factors might be extraordinarily high and shortly thereafter aircraft could be flying almost empty. Most marketers find it difficult to ignore high load factors but do not find the latter unacceptable. This often results in excess capacity.

Too many seats may come into a market for a variety of reasons such as aircraft positioning on a route, utilization targets, market share aspirations, or long-range objectives of a carrier. The market environment can change quickly without warning, leaving empty seats and rising costs.

Pricing policies are made by humans, and cutthroat fares can throw a market pair into turmoil. Most often this fare slashing is designed to gain share of the market or for reasons other than profit. Experience has shown that the results can be devastating to all airlines on the route. The industry has been vigorously pursuing various pricing experiments with little success. Cracks are appearing as a result and more and more airlines are repricing on a fully allocated basis.

Pricing experiments as marketing ploys have been both utter failures and smashing successes ever since deregulation, and many carriers turn to price promotions whenever they need quick cash.

Many carriers, however, have found it difficult to generate funds internally for new equipment on a large scale, or even to borrow from financial institutions. A gradual transformation is taking place as the industry carefully assesses its financial position and requirements for stability to remain viable over the long term.

OPERATIONS

An operational analysis examines all aircraft candidates and, although seldom makes the decision, serves an important role in the elimination of potential aircraft. Typically, only two or three aircraft are under consideration for a particular type of fleet expansion.

For example, it is known whether long-range or short-range aircraft will serve the routes. Two-engine aircraft have had limitations if the route is over long distances of water. If the traffic demand is at the cusp between two- or three-engine aircraft, the operations analysis will help to make a decision. In general, however, management has selected two or three aircraft models and fleet planners are directed to evaluate them.

Flight operations is concerned with the flight characteristics of the aircraft. Commonality with the rest of the fleet in cockpit layout, instrumentation, and avionics is important. Flight training costs are investigated. Training might either be done in-house or at another airline's facility. All of these factors have a cost impact and are an integral part of aircraft evaluation.

Technical operations include aircraft and engines in both overhaul and maintenance. The level of spares required is a cost consideration. If there is fleet commonality, there should be no additional costs in training. If the aircraft under study are different in engine or airframe, an extensive training program may be required. This matter of fleet compatibility is a very serious one.

Several years ago, Turkish Airlines (THY) ordered Fokker aircraft. However, the company found that it could not get delivery as rapidly as it wanted, so THY placed a concurrent order for several Fairchild F-27 aircraft produced in the United States. It was a shock when both types of aircraft were delivered; the Fokker was in the metric system and the Fairchild was in the American system of measurement. Maintenance, tooling, and many spare parts had to be duplicated for each system. Philippine Air Lines bought surplus DC-8's from Delta Air Lines (DAL) but did not buy the spare parts inventory. It was later found that although components retained the same part number, many had been modified according to DAL specifications and could not be replaced with standard DC-8 parts.

For many years TransAmerica Airlines (Oakland, California) had a maintenance contract with its neighbor and competitor, World Airways. The company's strategy was to limit the capital investment and maintain the flexibility to change its fleet without enormous maintenance training costs and parts support. The trade-off was increased hourly maintenance costs vs. lower capital investment and increased flexibility.

Direct operating costs (DOC) may be more important than the aircraft purchase price in some cases. Because aircraft are operated anywhere from 5 to 15 hours a day, DOC is a major consideration. It was said, in the early 1980s, that the DC-10 had a DOC about 5% to 10% lower than the L-1011. Assuming a DOC of $2,500 per hour and operating 10 hours a day, this could mean $25,000 per day or a $750,000 per month differential, not far from the debt service required to buy the aircraft.

At the top of this scale in utilization are the majors, flying their aircraft about 10 hours per day. The regionals and commuters experience block-hour utilization closer to five hours a day. The majors fly long-haul segments that allow them to achieve double-digit daily utility. Regionals and commuters often make as many as 10 or more landings (cycles) on short route lengths. The highest costs are in the takeoff and landing regimes. The lowest DOC and fuel economies are at high cruising altitudes, typically about 12,000 meters (about 37,000 feet) for jets.

In addition, regionals and commuters serve less dense and smaller airports; and the approach, holding, and taxi times are less. When majors land at Kennedy Airport or O'Hare, it is not unusual to taxi three or four miles to a gate, taking 15 to 20 minutes of ground time.

Fuel and crew costs have multiplied in the last 10 years and can represent as much as 60% to 70% of DOC. These figures vary owing to airline labor and fuel contracts, as well as efficiencies and the operating environment.

Despite economy measures adopted by air carriers and dramatic improvements in aircraft management techniques, present-generation aircraft have reached the outer limits of fuel economies. Even the Boeing 757 (186 passengers) is said to use only 15% less fuel than the Boeing 727-200 (143 passengers). A used 727 can be leased for about one-tenth the cost of a new B-757.

Therefore, the question that must be resolved by the equipment selection team (EST) is whether the 15% savings will support the increased debt load.

It explains why aircraft declared as obsolete have found new leases on life with start-up carriers. TWA declared its DC-9-10 aircraft to be surplus and fuel inefficient, yet Midway took over the aircraft and operated it to reach the highest return on investment in the industry in its second year. Midway was profitable in the fifth quarter while TWA posted enormous losses for the same time period. United has sold almost all of its 727-100 aircraft. Northeastern International Airlines (NIA), operating from Long Island to Florida, bought obsolete DC-8-52 airliners. With an investment of $50,000 in initial funding, NIA reached revenues of $50 million in its *second* year of existence. The airline grew rapidly, and managing this growth proved to be NIA's nemesis.

Declaring aircraft obsolete is fraught with considerable uncertainty and requires careful study and evaluation.

Range and growth are important factors. Milton Friedman, the Nobel economist, nobly observed that "there is no free lunch." This statement applies to equipment selection. If a carrier buys an aircraft with range in excess of need, there will be increased operating costs due to a heavier aircraft with the extra fuel tanks. Landing fees are usually set according to the maximum landing gross weight (MLGW), and a longer range more often results in a higher MLGW; hence, higher landing fees.

Because of the economies of fleet commonality, the EST must examine the derivative growth of each aircraft. The Boeing 737 was offered as a 100 model, then a 200, and now a 300. Southwest Airlines operates all of these aircraft, and each one serves a slightly different market more efficiently, providing economies of commonality in spares, engines, crew training, and operation, as well as scheduling for block times and capacity.

An unfortunate selection for a major carrier was the Lockheed L-1011-500, a highly sophisticated trijet. It proved to be the last one of the line and, as a result, the next equipment upgrade had to look elsewhere. Many airlines turned down the 500 model for this reason, with the result that the "de-bugging" of that model was limited to just a few carriers. With such limited feedback, Lockheed was unable to impress enough carriers that the L-1011-500 was a good risk.

This R&D was an important consideration, for the Boeing 727 was owned and operated by more airlines than any other aircraft. Customers were constantly feeding Boeing with incremental improvements, which were then disseminated to all operators, increasing efficiencies, heightening passenger acceptance, and generating new sales of aircraft.

FINANCE

Capital requirements for the industry for new or replacement aircraft could be over $100 billion over the next 10 years. Two primary characteristics determine a carrier's ability to secure capital investment: aircraft acquisition and traffic growth.

Unfortunately, the capital demands of the world's airlines have been surging ahead of their rate of capital accumulation. In the past, the gap has been bridged by borrowing, leasing, subsidy, and appreciation of purchased equipment. Borrowing and loan subsidies will probably continue in a limited manner as will appreciation and depreciation of equipment for tax purposes. However, there is increased competition of capital resources, greater debt burdens, and the imposition of highly restrictive credit terms.

That was a problem harassing TWA as the airline was the only losing corporation within its holding company. Borrowing provided funds for the more profitable firms in the holding company, and the loss-leader TWA was unable to compete for funds for growth and expansion. Finally, the carrier was spun off to go out into the world on its own, and is now doing much better under the leadership of Carl Icahn.

A good example was when Eastern Air Lines signed a labor contract giving its unions substantial benefits. Eastern's investment bankers demanded that the contract gains be modified or they would not provide additional funds.

In evaluating a purchase, the EST prepares a statement of anticipated earnings, costs, ROI (return on investment), debt service, source and use of funds, cash flows, risk analysis, a current and realistic balance sheet, and opportunity costs. The output of each aircraft type must be modified to a comparable basis, apples to apples, to accomplish the above aims.

Often, when analyses have been completed, aircraft under study appear equal. This is not unusual because advances in aircraft technology are shared to a high degree by manufacturers, and airlines are queried about their needs well in advance of the basic design. This has allowed for a colinearity and a financial quantification procedure to provide data for the selection of equipment, but falls short of rendering a decision.

It is in the final step of the *contract negotiations* that the decision is made. There are rarely fixed numbers in prices and conditions when it comes to aircraft sales. The costs of aircraft are highly elastic depending on competition for the airline's resources, numbers of aircraft purchased, down payment, integrity of the firm, the contract period, and many other considerations.

Within this framework, less than optimal aircraft (for that carrier) may get the go-ahead because of management biases, better warranty offerings, training, inventory stocks, short-term benefits, advance payment, and a financial package tailored to the needs of the carrier. The obvious equipment choice may be relegated to a secondary position simply because of the negotiated benefits.

Eastern Air Lines did not want an aircraft as large as the Airbus A-300B. It was sold to the company at the capacity price Eastern required. Eastern needed about a 185-passenger aircraft and not the over-200-passenger A-300B. Airbus Industries was intent on penetrating the U.S. markets and was able to sell the Airbus A-300B to Eastern because the debt service was ingenuously constructed as if the capacity of the aircraft was what Eastern wanted. When Eastern's load factors increased to support the actual size of the Airbus, payments were to increase.

Another example was the DC-9-80, which was not selling. McDonnell Douglas came up with the plan to sell it for nothing down and just debt service on the purchase price for a period of time. It was a clever financing scheme that solved the purchase problem for both carriers and the manufacturer. Many carriers, including American Airlines, took advantage of the offer.

SUMMARY

Aircraft, ground-support systems, and spares inventories can account for up to 80% to 90% of an airline's fixed assets. Approximately 50% of the assets are represented by the flight equipment with the balance split between ground support and spares. Additionally, the characteristics of the aircraft (new or used, large or small) dictate the operating procedures and operating costs, determine the airline's image, limit markets, and are important influences on passenger acceptance. Thus the equipment selection decision is a critical one.

Having been given the operating environment and conditions under which the aircraft will be competing, the task of the EST is an intricate one and provides management with a recommended acquisition plan. How management perceives growth and strategic planning will determine, in large part, the optimal choice.

The financing package will either support or alter the equipment selection. Prior to World War II, an airline's financing needs for equipment purchase were met primarily through the sale of stock. Following the war and up to the time of deregulation, long-term borrowing became necessary to finance major fleet expansions.

Some equity financing was also conducted in the mid-1950s when carrier earnings and stock prices were up, but subsequently the capital requirements far exceeded retained earnings, and equity financing was the exception and not the rule.

With the introduction of jets, virtually all new aircraft were purchased through debt instruments. Institutions financing aircraft purchases were mainly concerned with the practical life of the equipment and possible uses by other carriers in the event of a default.

The equipment selection process is straightforward, but the factors interacting are unpredictable and inconsistent. Planning and scheduling describe the requirements and constraints in response to the marketing environment. After the aircraft has been typed, operations inputs its evaluation of equipment and additional costs, then finance establishes its guidelines and negotiates the contract, working closely with marketing, operations, and management to reach the final equipment selection.

Chapter 9

Finance

Courtesy of Airbus Industrie of North America; Airbus A 300-600

DESCRIPTION

The role of an airline's financial manager has changed dramatically over the years. The manager's responsibilities have broadened and become increasingly important to the carrier's continued existence. Prior to the 1950s, the financial manager was mainly responsible for keeping accurate financial records, meeting expenses, and obtaining short-term and long-term funds when needed. Aircraft acquisitions were financed through the sale of stock.

During the 1950s, however, a greater emphasis was placed on controlling the flow of internal funds and on capital budgeting. Neither of these areas had been emphasized previously. Moreover, as commercial jet transports were introduced, airlines found themselves spending three times as much for a single aircraft as they had ever done before. Because the capital requirements far exceeded their retained earnings, airlines had to use debt instruments to finance their purchases.

Since that time, technological innovation has increased and the industry has grown tremendously. Those same responsibilities that the financial manager had in the 1950s became more complex, larger in scope, and more influential.

When deregulation took place in 1978, the role of finance in the industry increased severalfold. The financial manager, no longer to make public much of the reporting of performance, had greater responsibility and his (or her) influence began to touch all facets of the airline operation.

FUNCTIONS

In general, the financial manager's responsibilities can be categorized as either "internal" or "external" financial functions. The former involves maintaining efficient operations and accountability, while the latter pertains to financial planning, capital management, and meeting capital needs.

In the following sections, responsibilities, as well as the influence of government participation, will be discussed. An overview of airline financial statements and financial statement analysis will also be presented.

Internal Finance

Internal finance generally promotes operational efficiency and maintains a reliable internal accounting system. The more important responsibilities include budget formulation and control, accounting, and record keeping.

A financial budget is a tool that management uses for planning and controlling expenditures for specific purposes. The budgeting process is a continuous attempt to specify what should be spent to complete the job in the best possible way. After the budget is formulated through the determination of performance standards (which are periodically adjusted to reflect environmental changes), the actual results are continuously compared with the budget. This comparison serves as a basis for evaluation. A good understanding of an airline's financial relationships is essential when setting performance standards. Unreachable goals may result in resentment and frustration. Conversely, if the standards are too low, costs may be out of control and profits will diminish. A feasible budgeting system should play a positive role in the airline, improving internal operations and, hence, minimizing costs and increasing profitability.

The accounting and record-keeping function identifies, measures, records, and classifies the activities of the airline. These data are interpreted and summarized to formulate the airline's annual statements.

Thus, the underlying accounting system must be reliable and have suitable internal controls to ensure complete and accurate data. An airline usually has an internal audit department that continuously evaluates the accounting system.

External Finance

External finance functions include financial planning, cash management, short-term financing, and long-term financing. Decisions in these areas directly influence the overall development and profitability of the airline. The size of the airline, profits from operations, liquidity, business risk, financial charges, and financial risk are determined by top management.

The financial planning function can be categorized according to both the cash budgeting and capital budgeting processes.

Cash budgeting is the principal tool used in making short-term projections of future cash receipts and cash disbursements over various time intervals, such as a day, week, month, or quarter. The financial manager should make certain that there is enough cash available to meet current needs and that any excess funds are invested.

With the cash budgeting process, the finance manager can plan for the short-term financing of these needs, and also plan for the short-term investing of the temporary surplus of funds. Thus, the manager exercises control over the cash and liquidity of the airline.

Capital budgeting refers to making decisions involving capital allocation to various investment proposals. For an airline, this decision primarily concerns the addition to or replacement of aircraft in the present fleet. The capital budget process is complex; the financial manager must evaluate the financial ramifications of the proposals, estimating future revenues and costs, by using various analytical tools: cash flow analysis, profitability, return on investment, break-even analysis, pro forma balance sheet and income statement, and required financing. The importance of having reliable data to make the estimates cannot be overemphasized, for the analysis is only as good as the data on which it is based.

Managing cash involves maintaining cash availability and attaining maximum interest income on any excess funds. Optimal levels of both cash and short-term marketable securities should be determined. These decisions are based on many factors, including predictability of future cash streams, market conditions with respect to short-term interest rates, and so on. Cash budgeting, discussed earlier, is instrumental in carrying out these responsibilities.

The efficiency of cash management can be improved through various collection and disbursement methods that result in the maximum availability of funds. The basic idea is to accelerate accounts receivable collection, while delaying accounts payable disbursement. Collections can be expedited primarily by means of concentration banking and a lock-box system; disbursements can be slowed by "playing the float" and through the use of drafts. The delay in disbursements should not, of course, endanger a carrier's good credit standing with suppliers, since one firm's loss may be another firm's gain.

There are a number of short-term marketable securities in which an airline can invest idle funds. Those individuals making the decision should consider the default risk, marketability, yield, maturity, and taxability of the investment alternatives. Although the factors are not mutually exclusive, they are intended to serve as a basis for comparison.

Figure 9.1 shows the short-term investments of American Airlines as of December 31, 1988 and December 31, 1987, as listed in the notes to the 1988 financial statement.

If the future cash flows are known with reasonable certainty, the portfolio can be arranged such that security maturities coincide with when the funds will be needed. Consequently, large amounts of securities need to be sold unexpectedly, and investment in risky securities can be increased, maximizing the return on the portfolio. On the

	December 31	
	1988	*1987*
Cash	$ 54.9	$ 44.8
Tax Exempt Securities	—	276.0
Certificates of Deposit	859.6	380.0
Prime Commercial Paper	328.4	289.3
Other Short-Term Investments	43.7	22.3
AA Annual Statement	$1,286.6	$1,012.4

Short-term investments are carried at cost, which approximates market value. Thus, a portfolio of these marketable securities can be selected and managed.

Figure 9.1 American Airlines: Cash and Short-Term Investments (in millions of $s)

other hand, if the future cash flows are largely uncertain, the securities' marketability and market value fluctuation (risk) should be primary considerations. The return on the portfolio will generally be lower because of (1) the portfolio being composed of a good percentage of low-risk securities (typically Treasury securities), and (2) the additional transaction costs incurred when securities are sold unexpectedly.

SHORT-TERM FINANCING

Short-term financing is defined as credit that is scheduled to be repaid within one year. There are occasions when the airline needs short-term credit to meet current cash needs. The major sources of short-term financing for airlines are trade credit, accrual accounts, and commercial banks.

Trade credit refers to goods bought from other companies on credit and recorded as a liability under accounts payable. It is convenient, informal, arises from the normal business transactions of the airline, and, as such, represents a continuous form of credit. Trade credit, however, is only a discretionary source of short-term financing if the airline has some flexibility with respect to the payback period. "Stretching" accounts payable means paying a bill after its due period. The opportunity cost of "stretching" is the possible decline in the airline's credit rating.

For example, suppose an airline purchases on credit $500,000 worth of spare parts, materials, and supplies related to flight equipment from a supplier every four months. The supplier does not offer a cash discount and would like to be paid within 30 days. If the airline typically takes 45 days to pay this bill, the supplier, in essence, is extending credit of $500,000 for 45 days three times a year to the airline.

If a supplier offers a cash discount, this discount represents an opportunity cost if not taken. In this situation, trade credit, as a form of short-term financing, can be expensive.

Accrual accounts, like trade credit, represent a continuous source of credit. Wages and taxes are the most common accrual accounts. They are much less of a discretionary source of financing than is trade credit, since wages and taxes have to be paid on established dates. The only flexibility the airline has is the frequency of paydays for employees. This can be lengthened, within a very narrow range, to increase the accrued wages account.

Commercial banks are the most discretionary source of short-term financing, offering conventional promissory notes. Typically, short-term loans are "self-liquidating" in that they are repaid within a year. However, if at maturity the borrower is not able to repay a loan, the debt is usually refinanced.

The terms of a loan are determined through personal negotiations between the borrower and the bank. The interest rate charged varies depending on the borrower's credit-worthiness, the profitability of the borrower's relationship with the bank, and the existing prime rate. The borrower is often required to maintain a compensating balance; namely a demand-deposit balance equal to 10% to 20% of the funds borrowed.

A loan can be either *unsecured* without collateral, or *secured* by the borrower putting up collateral. For airlines, short-term investments and accounts receivable are usually pledged as collateral on short-term loans. Unsecured loans are extended under a line credit, revolving credit, or on a transaction basis to the bank's financially sound borrowers. On the other hand, unproven borrowers, or borrowers regarded as not financially strong enough to merit an unsecured loan, are required to put up collateral to reduce the bank's risk of loss.

Protective covenants, contained in loan agreements, are financial restrictions imposed on the borrower. Such restrictions serve to protect the bank, for if any of the covenants are violated, the bank can step in and take action, even demanding immediate and full payment. In most situations, however, the only action the bank takes is to work with the borrower and help with the borrower's problems. Common covenants found in short-term loan agreements include a minimum current ratio, restriction on dividend, and similar provisions.

LONG-TERM FINANCING

Obtaining long-term financing is important to the airline industry since the industry is highly capital-intensive. For example, equipment purchase price alone is substantial, running into tens of millions of dollars. A Boeing 747-400 jumbo jet costs approximately $120 million in 1988.

New aircraft must be continually purchased, both to replace older equipment in the existing fleet, and to provide the additional capacity that the growing airline industry demands. Normally, a company uses long-term financing methods to finance long-term assets, and the airline companies are no different. To finance an equipment purchase, the financial manager can select from many long-term financing alternatives, each with its own advantages and disadvantages.

The manager should try to obtain the mix of financing or capital structure that maximizes the valuation of the airline. Other considerations should be the airline's ability to meet any fixed financial charges; the timing of debt or equity issue; and the flexibility offered by each financing alternative concerning which financing options will be still available or closed in the future.

The alternatives include equity financing, long-term loans, bonds, leases, lease purchases, equipment trust certificates, conditional sales contracts, or the government. We will discuss each of these in turn below (except for government, which will be covered in the section titled "Government Participation"). After reviewing the alternatives, we will touch on the projected capital requirements and present financial position of the airline industry, along with their implications regarding future financing methods.

Equity financing includes retained earnings and the sale of common stock or preferred stock. A general and important benefit accruing to the airline in using equity to finance equipment purchases is that the airline retains ownership of the equipment. Consequently, the airline can take advantage of any investment tax credit and depreciation benefits, thus lowering the airline's overall liability.

Retained earnings result from positive net income being kept in the airline. Equipment financing through this source involves the least amount of corporate risk and no dilution of shareholders' equity. An airline typically uses retained earnings in combination with other financing methods, because the internally generated cash funds are not sufficient to meet the capital requirements.

Selling common stock raises funds through the capital market. The prevailing market price is an important determinant of the amount of stock that would have to be sold to raise the necessary capital. One advantage of using this source of funds is that it entails very little risk for the airline; the company is not obligated to pay out any dividends and there is no fixed maturity date. Another advantage is that selling stock increases the company's credit-worthiness, since creditors will have an additional cushion against losses.

The primary disadvantage of selling common stock is the dilution of shareholders' equity. By selling additional shares of stock, the existing shareholders will have less control and profits will have to be shared with more owners. Shareholders' dilution became a hot issue in the Texas International (TI) acquisition of Continental Air Lines (CO).

The employees felt that the economies of scale would cause wholesale layoffs at CO so they determined that additional stock should be issued and the employees would buy the shares out of future earnings. Many shareholders held the view that the takeover by TI might provide a more desirable alternative than the dilution of the stock's net worth by as much as 50%. However, the employees were not able to gain control of Continental, TI completed the takeover, and many employees were laid off.

Selling preferred stock also raises funds through the capital market. Preferred stock is a hybrid form of security that has the characteristics of both common stocks and bonds. Generally, preferred stocks carry a stipulated dividend, similar to bond interest.

However, while failure to pay bond interest results in default of the obligation, this does not occur if dividends cannot be paid. Nor does failure to pay the dividend result in company insolvency (in this regard, preferred stock behaves like common stock). Virtually all preferred stock issues have a call provision that gives the airline the option to "recall" the outstanding preferred stock at a stated price. One advantage of utilizing preferred stock is that there is no shareholder dilution. Existing shareholders retain their control and earnings potential. Another advantage is the flexibility that the call feature offers the airline.

The major disadvantage is cost. Preferred stock must be sold with a higher yield than bonds in order to compensate the investor for the additional risk, and the dividend payout from these stocks (unlike the payout from bonds) is not tax deductible. Although there is no legal obligation on the part of the airline to pay the dividends, many companies do regard the obligation as fixed.

Convertible preferred stocks are also issued, which give preferred stockholders the option to convert their shares into a specified number of shares of common stock. The advantages and disadvantages of this type of stock are discussed in the section on convertible bonds later in this chapter.

Long-term loans are promissory notes with maturities over one year. Sources include commercial banks, insurance companies, and equipment suppliers. These lenders look at the future cash flows of the airline because the debt, refinanced or not, eventually has to be paid back with earnings.

The terms of a loan are decided through negotiations between the borrower and lender. Loans can either be unsecured (without collateral), or secured with collateral, typically the aircraft. Moreover, there are restrictive covenants that serve to protect the lender.

General provisions usually used are a working capital requirement, cash dividend and repurchase-of-stock restriction, capital-expenditure limitation, and a limitation on other indebtedness. Interest rates can either be fixed or tied to the prime rate.

One advantage of using long-term loans for equipment financing is that the airline retains ownership of the equipment, which allows it to be the beneficiary of the investment credit and depreciation. Other advantages are that interest payments are tax deductible and only shareholders participate in superior profits when earned, since loan payments are fixed charges.

One disadvantage is the increased risk to the airline because interest is a fixed charge that might not be paid during unprofitable periods. Another disadvantage is that the covenants protecting the lender are usually very restrictive. Finally, provisions must be made to retire the debt.

Bonds, which are similar to long-term loans, are another financing alternative. The coupon rate of a bond parallels the interest rate of a loan; the airline deals with a trustee who is the representative of all bondholders, comparable to the lending institution in a loan agreement. The terms of the bond issue, included in the bond's indenture, are similar to the terms contained in a loan agreement. Nearly all bond issues have a call provision that gives the airline the option to buy back the bonds at a stated price.

There are three types of bonds: debenture (unsecured), mortgaged (secured), and convertible. The features of the debenture and mortgaged bonds, as well as their advantages and disadvantages, are similar to those discussed earlier for unsecured and secured loans, respectively (see above).

Convertible bonds, like convertible preferred stock, give convertible bondholders the option to convert their securities into common stock under specified terms and conditions. The conversion ratio, which stipulates the amount of common stock the convertible securities holder will receive at conversion, and the conversion price, which is the effective price paid for the common stock at conversion, are established when the bond is sold. The conversion price is set higher than the prevailing market price at the time of issue.

The advantages of selling convertible bonds or convertible preferred stock include lower coupon rates and dividends, respectively. Furthermore, if conversion occurs, the airline will essentially have eliminated these obligations and raised capital at higher stock prices than those that existed at the time of issuance. Investors are willing to accept lower coupon rates or dividends to own this conversion privilege.

Disadvantages include those related to a bond or preferred stock by nature, and the possible dilution of shareholders' equity. If conversion takes place, the current shareholders will have less control and a lower percentage of the profits.

Leases are financed by three primary types of arrangements: sale and lease-back, operating lease, and leveraged lease. These arrangements differ in certain respects, but they are also similar in that, in each case, the airline (lessee) operates the aircraft and makes periodic lease payments to the lessor, who retains title to the aircraft. Thus, the lessor, not the airline, takes advantage of the investment tax credits and the tax shelter arising from the equipment depreciation.

Financial institutions, because of their profitability and high tax brackets, have been attracted to these substantial tax benefits and have become the predominant lessors. Other lessors include insurance companies, leasing companies, and aircraft manufacturers (see discussion below on conditional sales contracts). The airlines do realize a portion of the benefits accruing to the lessor, however, in the form of lower lease payments.

For tax purposes, the lease payments are generally tax deductible. Finally, by leasing and not borrowing funds, the airline maintains its borrowing capacity.

In a sale and lease-back arrangement, the airline sells a plane it owns to another party (lessor), and this party leases the equipment back to the airline. As a result, the airline receives the cash from the sale and use of the equipment for the leasing period, but must make periodic lease payments and give up the title to the equipment. The lease payments are set up such that the lessor receives the full purchase price of the equipment plus a stated return on investment. The airline can make arrangements with the lessor concerning the option to purchase the equipment at the end of the lease period.

In an operating lease, the lessor buys the equipment from the aircraft manufacturer or from another seller and, in turn, leases the equipment to the airline. Except for this characteristic, a financial lease is similar to a sale and lease-back arrangement.

Leveraged leases involve the airline, the lessor, and a third party, designated as the lender. As far as the airline is concerned, a leveraged lease resembles any other lease. For the lessor, however, the situation differs from other lease arrangements. In purchasing the equipment, the lessor makes an equity investment, say 20%, and borrows the remaining 80% from the lender. The loan is secured by the equipment, by the assignment of the lease and lease payments, and occasionally by a guarantee by the airline itself. Lenders are often limited partnerships, financial divisions of the manufacturer, banks, governments, and other organizations interested in both the income and tax benefits.

Equipment trust certificates are a form of lease financing, and also represent a long-term fixed income investment to the buyers of the certificates. When the aircraft is delivered to the airline, equipment trust certificates are typically sold to institutional investors. The proceeds of the sale are used to pay for the aircraft. The title to the equipment is held by a trustee, who leases the aircraft to the airline. The trustee uses the lease payments to retire the certificates, also paying a fixed return to those certificates that are still outstanding. After the last lease payment has been made, the title of the equipment passes to the airline.

Equipment trust financing not only shares the advantages and disadvantages mentioned in the discussion concerning leases, but also the benefits, since equipment trust certificates enjoy a high standing as long-term fixed-income investments.

According to the terms of a conditional sales contract, the airline operates the plane, but the aircraft manufacturer retains the title to the aircraft until the airline has satisfied all of the terms of the contract. Under this plan, the airline agrees to make a down payment and subsequent periodic installment payments until the purchase price has been paid. After the last payment has been made, the airline receives title to the plane. If the airline does not satisfy the contract terms, the aircraft manufacturer can repossess the equipment.

The advantages and disadvantages of this type of financing are similar to those stated in the discussion concerning leases. The airline typically pays a lower effective interest rate on the installment payments than if it had purchased the equipment outright through long-term loans or bonds. However, in not owning the equipment, it loses certain tax benefits it would otherwise have received.

Because of expected continued growth in passenger and freight traffic, the projected capital requirements of the airline industry in the next 10 years are substantial, estimated to be nearly $100 billion. This estimate is about eight-plus times the capital invested in the 1960s, and six times that invested in the 1970s (see Figure 9.2).

CAPITAL INVESTMENT
U.S. SCHEDULED AIRLINES
1960 – 1990

$ BILLIONS

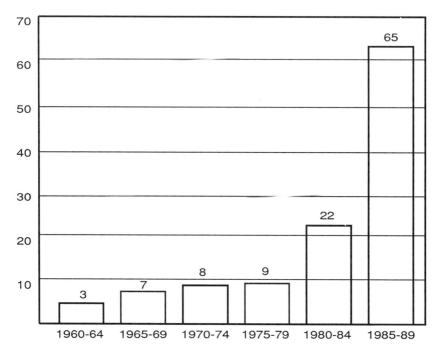

Figure 9.2   Industry Capital Requirements

The financial position of the American airline industry is characterized by more than twice as much financial liability as equity, and little profitability in the last few years. In 1981, U.S. scheduled airlines lost $300,826,000. This is an operating profit margin of (1.22), and a return on investment (ROI) of 5.3%. In 1982, the industry increased its losses to $915,814,000, showing an operating profit margin of (2%). The ROI was 2.7%, less than the troubled railroad industry and a fraction of most industries in the United States.

Judging from the airline industry's high leverage, below-average profitability, and large financing requirements, it is likely that airlines will be financed largely by convertible securities, equipment trust certificates, and leases. The requirement for earnings to support new equipment purchases is about 8% ROI, not probable in the near future.

As to using these financial vehicles, McDonnell Douglas offered the MD-80 on a straight lease. The offering has been so successful that most of the majors have added the stretched and quiet DC-9 to their fleets. Both the convertible securities and equipment trust certificates have been implemented. All three methods, in general, have lower effective interest premiums than do long-term loans, nonconvertible bonds, and nonconvertible preferred stock, and do not increase the leverage of the airline appreciably.

Financing through the sale of common stock has gained strength in the last few years; not on the basis of the industry's poor earnings but because of the future prospects. The glamour of the industry persists and the horizon does not look too dismal. Ever since deregulation, the airline industry has suffered unbelievable losses, but it seems to be finding an equilibrium between costs and revenues. A 1% increase in yield means about $300,000,000 in annual revenues to the industry, not an impossible achievement.

Also, several of the majors continue to post profits. The sale of common stock is limited as an opportunity because of shareholders' dilution and cannot be deemed much more than a one-time instrument in the current economic environment.

GOVERNMENT PARTICIPATION

The U.S. government, through monetary and fiscal policies influencing the overall level of economic activity, has greatly affected the profitability and financing of the airline industry.

Monetary policy refers to government actions that alter the availability and cost of available funds. For example, in periods of inflation, interest rates increase to compensate lenders for the decline in real purchasing power of their claims. The Federal Reserve System, in an effort to slow down inflation, tends to limit the supply of available funds. As a result, the competition for these limited funds intensifies, driving up interest rates even higher. Thus, an airline's cost of capital increases, the airline's capital requirements grow with the rise in prices, and available funds become even more difficult to obtain.

Foreign governments have also influenced the airline industry by serving as sources of funds for equipment purchases. These include various foreign export-import banks (such as the Export-Import Bank of Japan). The United States has a similar organization, called the Export-Import Bank of the USA (Exim Bank).

The Exim Bank only provides financing for foreign purchases of American-made equipment and is not available for U.S. purchasers. Some foreign governments, in an effort to secure aircraft orders for their national aircraft manufacturing industries, also have been either sources or backers of credit on very favorable terms.

A good example is the financial agreement between Pan American Airways and Rolls-Royce Limited, in connection with Pan Am's order of 12 Lockheed L-1011-500 aircraft in April 1978, with an option to purchase 14 more. The agreement covered $520 million of financing, repayable over 15 years. Because the order involved the purchase of Rolls-Royce RB211 engines, Pan Am was extended $120 million of United Kingdom export credit financing, with the $400 million remaining balance being placed with banks and institutional lenders, guaranteed by the United Kingdom's Export Credits Guarantee Department, and secured by the equipment. The guarantee reduced the financial risk substantially, and so lowered the effective interest rate below what Pan Am would have had to pay otherwise.

AIRLINE FINANCIAL STATEMENTS

Airline financial statements provide valuable information to the reader concerning a carrier's past performance, present condition, and future potential (estimated). This information is important to management, employees, shareholders, bankers, investors, and the government, all of whom are interested in the airline's financial position and performance. Many decisions are based on analysis of the financial statements. The principal statements are the balance sheet and income statement, accompanied by extensive footnotes.

If the airline is publicly held, other statements must be included, such as the statement of stockholder equity and statement of changes in financial position. In the next few pages, these statements will be discussed briefly, then followed by a short treatment of financial statement analysis.

The *balance sheet* reveals the airline's financial position at a particular point in time, and is reported as of the last day of the fiscal year, normally December 31. There are two sections in a balance sheet; the left-hand section lists the airline's assets, while the right-hand section lists the airline's liabilities and stockholders' equity. Both sections must be equal to each other because every asset dollar is claimed by either creditors (liabilities) or stockholders (owners' equity). Certain areas of the balance sheet appear atypical when compared to reports of companies in other industries.

One such area is the relatively small amount of inventories, which consist primarily of spare parts. This is not surprising because airlines deal with a "product" that cannot be inventoried, in the sense that an airplane flight cannot be stored in a warehouse.

Another area is the large proportion of assets that are fixed assets. For example, currently, property and equipment accounts for approximately 70% of the total assets of U.S. certified air carriers. Again, this is not surprising, since airlines are highly capital-asset-intensive.

The last area involves the right-hand side of the balance sheet. The airline industry, unlike many others, is highly leveraged, with about twice as much debt as equity.

The *income statement*, or *profit and loss sheet*, discloses the financial performance of the airline over the time period covered by the statement. Operating revenues, operating expenses, non-operating costs such as interest expense, taxes, and net income are reported. The *Statement of Stockholder Equity* typically shows the changes over the past year in retained earnings, common stock, additional paid-in capital, preferred stock, or Treasury stock. Statements differ among airlines as to which accounts are shown.

The *Statement of Cash Flow* examines the net changes in assets, liabilities, and equities, thus showing where cash came from and how it was used. This statement is one measurement of management's performance in utilizing funds. Since the airline industry is capital-intensive, one of the major sources of funds is depreciation and amortization. This is so because, although depreciation and amortization are annual charges against revenue, they do not involve cash outlays.

Financial statement analysis, in general, includes an evaluation of the airline's financial position and performance with respect to liquidity, leverage, activity, and profitability. The financial analyst needs "yardsticks" to make such evaluations. The most frequently used yardsticks are financial ratios.

Financial ratio analysis involves two comparisons: the comparison of the present ratios to the past and projected ratios of the airline under consideration; and the comparison of that airline's ratios to those of other airline firms.

Sources of industry ratio data include Dun & Bradstreet, Robert Morris Association (RMA), The Almanac of Business and Industrial Financial Ratios, trade associations such as IATA and Air Transport Association (ATA), and many individual firms such as financial brokerage houses, banks, and research firms.

Liquidity ratios are used to look at a company's ability to meet short-term obligations. The current ratio and quick or acid-test ratio are commonly used.

1. Current Ratio: $\dfrac{\text{Current Assets}}{\text{Current Liabilities}}$

2. Quick Ratio: $\dfrac{\text{Current Assets} - \text{Inventories}}{\text{Current Liabilities}}$

The current and quick ratios often seem excessively low in comparison to those of other industries. This is simply because airlines have small inventories (see discussion of balance sheet above).

Leverage ratios are used to measure and compare the levels of funds supplied by creditors and owners, and to look at the firm's ability to pay interest on its debt and other fixed charges. The higher the contribution made by creditors, the higher the leverage and risk of the firm. Ratios commonly used are debt to owner's equity; debt; times interest earned; and fixed-charge coverage.

1. Debt to Owner's Equity: $\dfrac{\text{Long-term Debt} + \text{Lease Obligations}}{\text{Owner's Equity}}$

2. Debt: $\dfrac{\text{Long-term Debt} + \text{Lease Obligations}}{\text{Long-term Debt} + \text{Lease Obligations} + \text{Owner's Equity}}$

3. Times Interest Earned: $\dfrac{\text{Profit Before Taxes and Interest Charges} + \text{Depreciations}}{\text{Interest Charges}}$

4. Fixed-Charge Coverage: $\dfrac{\text{Profit Before Taxes} + \text{Depreciations} + \text{Interest Charges} + \text{Lease Obligations}}{\text{Interest Charges} + \text{Lease Obligations}}$

The airline industry can be characterized by relatively high leverage and low coverage of fixed charges.

Activity ratios measure how well a firm employs its resources. Its inventory turnover and average collection period ratios are not very meaningful to airline financial analysis and will not be included here. Those ratios that are applicable are fixed-asset turnover and total asset turnover.

1. Fixed-Asset Turnover: $\dfrac{\text{Operating Revenue}}{\text{Fixed Assets}}$

2. Total Asset Turnover: $\dfrac{\text{Operating Revenue}}{\text{Total Assets}}$

These two ratios are characteristically low in the airline industry, given its large asset base.

Profitability ratios show the net result of the interaction of management policies and decisions, and they reflect the firm's earning power. Profit margin on sales; return on total assets or return on investment (ROI); and return on net worth or return on equity (ROE) are the ratios commonly used.

1. Profit Margin on Sales: $\dfrac{\text{Net Income}}{\text{Operating Revenue}}$

2. Return on Assets: $\dfrac{\text{Net Income}}{\text{Total Assets}}$

3. Return on Net Worth: $\dfrac{\text{Net Income}}{\text{Net Worth}}$

As noted previously, the airline industry has experienced low profit margin and return on assets. The return on net worth appears more respectable, but this is due to the fact that the airline industry is highly leveraged.

SUMMARY

Ever since the 1950s, the airline industry has experienced significant technological innovation and tremendous growth, two of the major reasons for the annual rise in the industry's capital requirements. Profits have not increased proportionately, however, and today we find the airline industry to be highly leveraged as well as capital-asset-intensive. In this scenario the responsibilities of the airline financial manager have become more complex, broader, and more crucial to the airline's well-being. The manager's responsibilities can be categorized as either "internal" or "external" finance functions.

External finance functions include financial planning, cash management, and obtaining both short-term and long-term financing. Capital is allocated to investment proposals and future cash flows are estimated to ensure that cash needs are met and any excess funds are invested. The goal of external financing is to build the best possible capital structure, one that maximizes the airline company's value, minimizes the financial risk, and allows for flexibility in the future. Thus, decisions in the external finance functional areas have a direct influence on the airline's future profitability and ability to compete.

Through its monetary and fiscal policies, the U.S. government influences the cost of capital and availability of both internally generated funds after taxes and available funds. Foreign governments have also begun to play an important role in providing financing for equipment purchases.

Finally, no discussion of finance is complete without some mention of financial statements and financial statement analysis. The important fiscal statements are the balance sheet and income statement, both of which provide useful information about the economic health of an airline.

Chapter 10

Determining the Financial Condition of an Airline

Courtesy of Boeing Canada, de Havilland Division; de Havilland Dash 8

DESCRIPTION

Financial analysis is designed to determine the relative strength of an airline. Is the firm financially sound and profitable vis-á-vis other firms? Is its position improving, deteriorating, or in jeopardy over time? Investors need such information to estimate both future cash flows from the airline and the stability of these flows. Managers need to be aware of their firm's financial position to sense and strengthen weaknesses in a continuing quest for improvement.*

It has often been said that publicly held airline companies construct at least three income statements and balance sheets for publication: one for the Securities and Exchange Commission (SEC); one for the shareholders; and, finally, one for internal use by management. Given the ability to modify the apparent financial condition of an airline, it is essential to be able to evaluate data and determine the financial condition of a carrier with some reasonableness. This chapter is a modified version of an excellent airline pilot negotiating document prepared by the Air Line Pilot Association (ALPA) staff, which has to be one of the clearest descriptions of how to find one's way through accounting mazes to determine the true current financial status and future expectations of an airline.

QUALITATIVE ANALYSIS

A *qualitative analysis* is a fundamental step in conducting a comprehensive assessment. The objective is to gain a broad perspective of the airline; its corporate structure, competitive position in the market, corporate personality and philosophy, management strategy, route structure, fleet composition and major changes in the airline over the past few years (managerial, structural, competitive, financial). The bulk of this general information is available publicly. The major information sources include the company's annual report, SEC filings (10K, 10Q, 8K, prospectus), Department of Transportation (DOT) filings, investment research reports, business and aviation periodicals, and articles from the press.

The SEC Form 10K generally provides the most comprehensive overview of the airline in one document. There are several parts to the 10K report, which is an annual SEC filing required by any publicly traded company.

Part I of the 10K document describes the company's overall struc-

*Material supplied through courtesy of Air Line Pilot Association Negotiation Department and the *Negotiator's Handbook*.

ture, lines of business, and subsidiaries. The form devotes considerable space to a discussion of each business segment. Topics included in the discussion are operations, market share, route structure, competitors, employees, management strategy, environmental consideration, real estate properties owned or leased, and the impact of legal proceedings on the company.

Parts II and IV of Form 10K are related primarily to financial and operational data. Included in this section are management's discussion and analysis of the financial condition and results of operations, the audited financial statements, the auditor's opinion, and disclosure of any disagreements with the auditors on transactions, management changes, security ownership by management, and current listings of executive officers and directors.

Annual Report

The *annual report* to stockholders will contain much the same information as the Form 10K, plus some additional information. Ordinarily, there is a discussion of operational highlights and a letter from management. Annual reports may contain information on recent changes to the company or future plans and strategies. The heart of the annual report is the audited financial statement.

Results for the past three years are generally displayed next to those of two or three prior years for easy comparison. A quick reading of the financial statements and the accompanying notes should be performed primarily to familiarize the reader with the earnings results for the past year. The reader should note from the income statement whether the business generated an operating income or loss for the year, and how this performance compared to the prior year. Income from operations is that income generated by the normal everyday business of transporting people and cargo. It is distinct from the net income figure, which includes gains or losses from activities other than normal operations, such as interest income, gains from sale of a subsidiary, and extraordinary events. While the reader should review the net income figure of the past couple of years, any unusual gains or losses that appear on the income statement should be noted for further investigation.

The balance sheet should be examined initially for major changes in assets and liabilities from the previous year, focusing particularly on increases in short- and long-term debt, the amount of stockholders' equity and retained earnings compared to the previous year, and the level of cash and cash equivalents. Negative figures for retained earnings or stockholders' equity represent real danger signals, particularly when accompanied by operating losses. To a large extent, an initial review of the financial statement should confirm much of what is already known about the company—that it is growing or contracting, that the company issued debt or equity during the year, or that the company sold a portion of its business.

A reading of the opinion of the outside auditors is essential. The reader should be on the lookout for "qualifiers" in the audit opinion, especially one that might indicate that the auditor questions the ability of the company to continue as a going concern.

The notes to the financial statement should also be regarded as required reading, since they serve an extremely important function. Notes to the financial statement provide explanations of accounting policies and disclosures of major fiscal events; those relating to long-term indebtedness provide descriptive information about the various notes and obligations as well as collateral arrangements and other loan covenants imposed by the lenders. The terms of the most recent loans provide valuable information about the comfort level of the bankers. The imposition of covenants that restrict operations in some way (for example, capital spending restrictions, requirements to maintain certain cash or net-worth levels, lender approval of top management changes) are red flags indicating lender concern about loan repayment.

The notes discussing employee benefit plans reveal the level of pension plan assets available to meet the actual computed pension obligations, any pension fund assets available to meet the actual computed pension obligations, any pension fund waivers that may have been granted by the IRS, and changes in actuarial assumption used to calculate future benefits. Notes covering aircraft purchase commitments and lease commitments indicate future equipment-related financial obligations. The analyst will want to determine that these and other commitments can be met.

So far, only information that has been prepared by the subject company has been examined, the bulk of which is devoted to the internal operations of the company. However, to conduct a thorough internal analysis, the reader must also focus on external factors that may impact the company's prospects and profitability. The reader should have a basic idea of general economic and industry forecasts, the position of major competitors, and other trends or issues facing the industry. Two primary sources for this type of information are investment analysts' reports and articles in the popular industry and financial press.

Investment analyst reports of specific companies present an independent and critical review of the company that cite both internal and external factors. They often compare one airline to its competitors to evaluate strength of operations, quality of management, and the appropriateness of strategies. Investment houses also issue industry forecasts of expected traffic and revenue growth, competitive analyses among the various industry segments (majors, nationals, regionals). It is useful to review several different analyses to gain different points of view.

Newspaper and magazine articles too often have a rather narrow focus, but occasionally one might be found that is more detailed. Feature articles in business magazines provide a good perspective on the overall well-being of the company. Through these articles, the reader can gain an understanding of the public perception of the company seen through the eyes of industry experts, suppliers, travel agents, and former employees. The reader is likely to find valuable evaluations of management and its actions or strategies, discussions of the position of the airline in the industry and anecdotal information that may be very telling. These sources are probably the best in terms of getting an insider's view of corporate goings-on.

Upon completion of a qualitative analysis, the reader should have a good feel for what areas need to be investigated more extensively. Additionally, the reader will have developed a broad understanding of the company and its history, which prepares one for the next stage of analysis, a quantitative review. A general knowledge of the company will add measurably in interpreting the quantitative results.

QUANTITATIVE ANALYSIS

Using the knowledge already gained, the analyst will begin a detailed quantitative analysis. The object of a quantitative analysis is to determine the financial strength of a company and its expected performance in the future. The sources of data used are the financial statements and operating statistics available from the DOT.

Balance Sheet Analysis

Often the analyst's first concern is the *liquidity* of the company. On the balance sheet can be found the total current liabilities, that is, debts that are due within one year from the date of the balance sheet. The source of funds from which to pay debts is current assets. Working capital, the difference between current assets and current liabilities, is the amount left free and clear if all current debt were to be paid off.

To judge the company's ability to meet its current obligations the analyst will calculate the *current ratio*. Normally, a company will wish to maintain a current ratio of 1:1, where current assets are equal to or greater than current liabilities; however, most airlines operate with a smaller ratio, because unearned ticket revenue (tickets sold but service not yet provided) is treated as a current liability. A ratio of current assets to current liabilities of 0.8:1 or less would be cause for concern particularly if accompanied by operating losses. If the current ratio appears low, the ratio should be compared to that of prior years to measure the extent of deterioration.

A poor liquidity position is correctable through debt restructuring whereby short-term debt or equity financing can be applied to pay off short-term obligations. An airline in a weak liquidity position will experience difficulty paying its day-to-day bills. Be on the lookout for reports of slow payment to suppliers or sudden corporate emphasis on cash-generating programs.

After investigating the company's liquidity, the analyst should evaluate the company's *capitalization.* Capitalization refers to the relative interest in the firm's assets of the two major sources of capital, the stockholders and the creditors. A capitalization scheme composed of long-term debt (debentures, equipment loans, capital leases) rather than equity can be considered highly leveraged. Highly leveraged companies represent a higher risk since they incur large fixed interest costs associated with high debt. Moreover, there is only a small amount of equity available to absorb losses.

The primary measure of leverage is the *debt/equity ratio.* A debt/equity ratio that increases over the years indicates that the company is taking on more debt to expand or maintain operations or, as a result of net losses, is shrinking its equity base. A debt/equity ratio of 2.5:1 or above should raise a red flag and is particularly worrisome if operating losses are anticipated. When operating losses are occurring, high-leveraged companies will find additional bank lending more scarce, if available at all, and very expensive.

Another area on the balance sheet to be reviewed is the airline's net worth or stockholders' equity. Net worth is the difference between total assets and total liabilities. It consists of funds received for the stock issued by the airline plus retained earnings. Retained earnings is the sum of all net profits or losses earned by the airline since inception, less any dividends. Should the *retained earning figure* become negative, the company will have begun to eat into its capital base. This is a danger signal that indicates the company has suffered real deterioration. If total stockholders' equity becomes negative, the company has exhausted its capital and the potential for bankruptcy is high. The analyst can determine if the company can sustain another year of losses by subtracting the current year's net loss from the remaining stockholders' equity. In this way, the analyst can assess whether the company can sustain another equivalent loss before equity is fully eliminated.

Statement of Cash Flow

The *Statement of Cash Flow* is a required portion of a financial statement and is also known as a *Sources and Uses of Funds Statement.* This statement details what funds or cash the company raised throughout the year and how these funds were spent. In the sources-of-funds section, the analyst would see the proceeds from the sale of an asset or proceeds from a bond or stock issue. Similarly, in the uses section one would see the application of fund to purchase new equipment or repay debt obligations. This is the closest that the analyst comes to obtaining cash-flow information lacking access to internal documents.

The analyst's attention is called to the Statement of Cash Flow because of a statistic that tells whether operations are providing cash or using cash. Look for a line entitled "Total sources of funds from operations," which is computed by taking net income or loss and then adding back any expenses that were deducted (or income that was added) in computing net income that did not require or provide cash. The simplest example is depreciation, which is deducted from revenues as an expense but not a cash outlay. When the sources of funds from operations is negative, this suggests that operations are a cash drain and that to sustain the operation, assets must be sold or additional capital raised to survive. A large negative statistic suggests that the company will have extreme difficulty meeting debt obligation payments. Negative statistics for two or more years are very clear danger signals. The analyst would find it worthwhile to compare the amount of cash provided from operations to the amount of debt obligations due within the following year.

Income Statement Analysis

While the balance sheet shows the fundamental soundness of a company, the income statement shows how well the company has done for the year. During the qualitative analysis, the analyst would have noted if revenues and expenses have been rising or falling. The key statistic on the income statement is operating profit or loss, which represents operating revenues less all operating expenses. A firm unable to generate an operating profit over two or more years is experiencing difficulty.

To evaluate operating profit in detail, the analyst will calculate the *operating margin* (operating profit divided by total revenue), which is the basic measure of profitability. By comparing the margin with that of previous years, the analyst can tell if the company has become more or less profitable. For example, the income statement may show that revenues and operating profit both fell in dollar amount from last year, but the margin percent is higher. This indicates that although revenues are off, the airline's operations were managed more efficiently. If margin comparisons over the past several years reveal a trend of decreasing margins, then costs are rising faster than revenues. This indicates less efficient operations and the causes should be investigated.

To determine which source of revenue is decreasing, or which expense is increasing relative to total revenue, the analyst will generally perform a vertical analysis. Each source of revenue and each expense is expressed as a percent of total operating revenue. When these results, such as crew wages as a percent of revenue, are compared to previous years, the items that have changed adversely will become apparent. Depending on the findings thus far, the analyst may feel a more detailed investigation of operating revenues and expenses is desirable.

The analyst might express the dollar change from one year to the next as a percent. Expressing the change in each source of revenue and each expense as a percent reveals relationships that might not be obvious from a review of the changes in dollar amounts. For example, an increase in fuel costs could be due to an increase in its price. But if the analyst knows that fuel prices have not risen recently, the analyst should check the airline's operating statistic for an increase in capacity. It may also be useful to express operating expenses in terms of block hours. This gives each item a common unit of measure to facilitate comparison of expenses per hour of operation. In the example above, if the increase in fuel costs was indeed linked to an increase in capacity, then the fuel per block hour should be relatively constant.

Below the operating profit on the income statement are the nonoperating income and expense items. The analyst should review the dollar magnitude and change in interest expense. A highly leveraged firm will incur high interest expense. To determine the company's ability to meet interest expense, the analyst would look at a ratio called *times interest earned* or *interest coverage*. This ratio measures the firm's performance in terms of how many times interest expense could be paid from earnings. If the ratio is less than 1, the firm is not earning enough to cover interest payments on debt obligations or debt principal payments. Creditors will want to know if previously borrowed funds have been put to good use and that earnings are sufficient to pay for the use of these funds before lending more.

Within the nonoperating section of the income statement can be found many other gain or loss items. The analyst will need to refer to the notes to the financial statements that describe gain or loss items. Often gains or losses are recognized that are nonrecurring. Examples include sales of subsidiaries, gains from sales of tax benefits, discontinued operations, accounting principle changes, and so forth. The "bottom line," while considered the ultimate measure of corporate performance and profitability, may be a less important measure than another figure: operating profit or loss minus the net of interest income and expense.

Analysis of Operations

Along with the financial data, there are nonfinancial operating statistics to be analyzed, much of which is unique to the airline industry and is useful for comparison to historical data. Also useful are comparisons to the operating statistics of competitors. The data can be found in the carrier's monthly operating reports and in periodic reports filed with the Department of Transportation (DOT) and the Air Transport Association (ATA).

An area of interest to the analyst is *capacity*. The standard measure of capacity is the *seat mile*, or one seat flown a distance of one mile. Seat miles are used as a standard unit of measure so comparisons can be made between equipment and different routes. A carrier's total capacity is expressed in *available seat miles* (ASMs) (all seats available for sale multiplied by the number of miles each is flown). It is important to compare total capacity for several years to determine if the company has increased or decreased its capacity. If available seat miles have increased and if the qualitative analysis showed no additions to flight equipment, then the company must be flying longer routes or more routes. To measure aircraft utilization, the analyst should examine average block hours per aircraft per day. An increase in capacity whether from more aircraft or better utilization of existing aircraft will not by itself improve revenues. Revenue depends on the load and yield—the amount of capacity that is used and the revenue generated per passenger mile.

A reader may wish to study capacity expenses in detail to determine how efficiently the carrier provides service relative to competitors. Capacity expenses are those expenses required to provide aircraft capacity, regardless of the degree to which that capacity is used. Capacity costs include flying operations expense, maintenance expense, depreciation and amortization, and aircraft ground-servicing expenses.

Operating expense per ASM measures the cost per seat mile of providing air transportation. On routes with a high degree of price competition, the carrier with the lowest operating expense per ASM will have the advantage of being able to earn an operating profit at a lower fare than its competitors. Another useful measure of flight operations expense is the block hour. Block hours are the sum of flight time plus taxi time. When analyzing operating expense variations, the focus is generally on the two largest expenses: salaries and fuel. Flight crew salaries per block hour and fuel costs have changed over time.

Traffic is measured in *revenue passenger miles* (RPMs), the number of revenue passengers multiplied by the number of miles each passenger is flown. Comparison of traffic figures over several years gives the analyst a feel for the effectiveness of the airline's strategy. The analyst must, however, be careful not to jump to conclusions in analyzing capacity and traffic changes. If the route structure has been changed significantly, if ASMs have remained the same, and if RPMs have fallen, then the route changes failed to increase traffic. However, the route restructuring may have been attempted to lower costs or increase revenues.

Perhaps the key statistic that can be used to measure the carrier's overall strength of operations is to compare the actual load factor to break-even load factor. *Load factor* is the ratio of revenue passenger miles to available seat miles. It represents the average percent of seats occupied by revenue-producing passengers. Decreasing load factors indicate that capacity is being utilized less than before. This may be caused by higher fares, increased competition, failure of marketing programs, excess capacity, seasonal variations, or a general economic downturn. Whatever the cause, the airline is carrying fewer passengers relative to available capacity, but still incurs the operating expenses leading to a decrease in operating income.

The break-even load factor represents the load factor that is required for passenger revenue to equal operating expenses, resulting in no operating profit or loss. Break-even load factor is sometimes computed on the combined interest expense and operating expense so as to factor in the cost of financing equipment purchases. An airline cannot operate successfully over a long term unless its actual load factor exceeds the break-even load factor. In today's deregulated environment, airline managements are setting goals to reduce the break-even load factor, focusing primarily on reductions to the operating cost component.

Perhaps the most oft-spoken operating statistic is *yield*. Yield equals the operating passenger revenue divided by RPMs. It is expressed in cents per RPM and indicates the amount one passenger pays to fly one mile. A change in yield may result from a combination of factors: change in fares, the number of passengers emplaned, or trip lengths. An important factor affecting yield is incremental pricing. Incremental pricing means selling a seat at a discount that otherwise would have gone empty. When used correctly, the carrier can obtain an additional contribution toward operating income. Unrestricted discount fares result in a decrease or dilution in yield and revenues.

ACCESS TO INTERNAL CORPORATE DATA

Financial analysis of an airline can be greatly enhanced with access to certain types of internal financial documents. While there is a natural inclination to gain access to "the books," the books and accounting records only convey information about what has already happened. The accounting records are historical documents that are summarized in the financial statement. A review of the financial statements will tell the analyst where the company has been and its current position. Given the opportunity to access company information, the reader should concentrate more on company projections and forecasts. This information provides insight into where management believes the company is headed and how it plans to get there. Access to internal corporate documents may also provide valuable nonpublic information about the company's banking relationships and pressures being imposed upon the carrier. Corporate documents that would prove to be very useful to the analyst include:

1. Projected Income Statement by month including supporting forecasts of load factors, available seat miles, revenue passenger miles, block hours, yield (forecast should at least be one year);
2. Projected Balance Sheet covering same period;
3. Detailed Cash Flow Projections covering same period;
4. Most recent unaudited financial statements with accompanying notes;
5. Schedule of debt and capital lease obligations complete with descriptions and payment schedule;
6. Copies of loan agreements including information of restrictive loan covenants;
7. Schedule of collateral supporting existing loans and listing of unpledged property—also current property appraisals;
8. Information regarding planned additions during upcoming year and planned asset disposition.

Presuming that the analyst has access to cash flow projections, it should be readily apparent when severe cash crunches are expected, the severity of a cash need, and whether the projected cash position will improve or deteriorate further. The projection should detail how cash deficits will be offset—such as borrowing against lines of credit, debit, or stock issue, wage reductions, or sale of assets. Cash flow projections are usually the result of income statement and balance sheet forecasts and should therefore be available. If the company is experiencing financial difficulty, it is likely that the firm's bankers will be requesting these forecasts, and attempts should be made to receive copies of those distributions.

In a review of the company forecasts, the first step will be to evaluate the company figures and particularly the underlying assumptions. For those months of the forecast where actual data are available, the figures in the forecast should be compared. The analyst may find that the forecast is rather optimistic and that certain assumptions should be changed. Using the company forecast as a basic model, the analyst can alter certain assumptions (for example, expected load, ASMs, yield, route changes) of the forecast and measure the impact of an assumption change on the carrier's profitability and cash flow. The availability of a microcomputer and a spreadsheet program greatly facilitates this "what if" sensitivity analysis. Several analyses or scenarios can be performed to examine "best case," "worst case," and "most likely" cases.

Several of the documents described above relate to corporate debt obligations. Here the reader is interested in the names of lenders, the amounts borrowed, the loan terms, the specific collateral supporting each loan, restrictions on future borrowing, and operational restrictions. The specific restrictions placed upon the company by the lenders will reveal how nervous they are about recovery of loans. For example, the analyst might learn that the company must maintain a minimum cash level or will be declared in default on the loan. Having available a cash flow to forecast, the analyst could determine if cash flow is anticipated to drop below the minimum level and in what month this would likely occur. As such, the analyst might be able to forecast the stress points throughout the year, presuming the forecast is a reasonable one. The analyst is also interested in whether assets exist that are not pledged as collateral and might be available to support additional borrowings.

If the analyst knows the market value of the carrier's equipment, as evidenced by an outside appraisal, the analyst will be able to determine how well protected the lenders are in the event of default, or if perhaps additional borrowing capacity is possible.

SIGNALS OF CORPORATE DISTRESS

When a company files for bankruptcy, it is seldom because of a single catastrophic event or without warning. Typically, there is a well-publicized series of indicators of distress leading up to the final announcement. Many of these signals are nonfinancial in nature and never appear in the financial statements. Analysts should be aware of these signals and make note of them as they progress through the financial statements, investment analysts' reports, and articles in industry and financial publications.

The first sign of deterioration consists of a general weakening in the company's financial position and its industry position. While there might not be any cash crisis, the company will have generally consumed a good share of excess cash resources and will have begun borrowing against credit lines. This action will be necessary because operations will require, rather than generate, cash resources. The reporting of an operating loss and a working capital deficiency would be expected at this stage. Operational performance in terms of load, break-even load, yield, and various operating costs per ASMs will likely show some deterioration and appear unfavorable when compared to competitors. Management will take steps to improve the profit performance by altering its marketing strategies and by perhaps initiating a cost-cutting program. At this stage, the company will attempt to maintain its employment force, its dividend policy, and its equipment purchase commitments. The prevailing view will be that the company is experiencing some problems that are correctable and that with a few changes the company can "get back on track."

If management is unable to correct the underlying causes of the carrier's problems, the company will reach a more advanced state of deterioration. Company actions will be of considerably greater urgency and severity than those associated with initial stages of stress. Actions might be taken that alter the basic financial and operational structure of the company. The prevailing view is that the company is in a "turn-around" condition. Specific internal actions or events that the analyst should watch for are:

- complete suspension of dividends
- loan defaults and loan restructurings
- sale of critical assets used in operations
- sudden resignation of members of the board or upper management
- search for merge partners

- deep discounting of fares on traditional routes
- appeals to the IRS for waivers of funding for pension plans
- use of new or unusual sources of financing
- beneficial adjustments to the financial statements
- approaches to labor units for concessions in compensation and productivity improvements
- pay cuts to management
- elimination of layers of middle management
- replacement of key management positions
- cash crises that result in exceptionally slow payment to suppliers
- cancellation of equipment purchase contracts

The actions and reactions of the banking community are important in measuring the bankers' level of discomfort, which is closely related to the overall financial condition of the carrier. At the initial stages of financial deterioration, the airline's bankers will begin to take steps to protect existing and future loans. Bankers may extend loans but only at high interest rates and sound collateral protection. Moreover, the bankers will add restrictive loan covenants, which are geared to limit actions of the company that would be unfavorable to the recoverability of their loans. The worse the company's financial situation, the tighter the control that bankers will attempt to impose. The analyst should watch for the following signals of deep distress:

- banks tightening covenants on loans by setting minimum levels of cash or net worth, operating targets, etc.
- severe difficulties gaining lender support in debt restructurings
- linking further extensions of credit to employees' wage concessions and other cost-saving measures
- demand for preapproval of upper-management replacements
- denial of additional credit at any price

Financial markets will begin to react at the first signs of financial stress:

- negative coverage in the financial press
- poor investment advisory reviews
- "sell" recommendations from brokers
- company stock in disfavor with institutional investors
- poor stock performance

If the firm's ability to survive becomes questionable, the carrier's bond rating will be downgraded to a speculative level. Outsiders will become aware of the advanced stages of distress as reports are published in the popular press. The signals include:

- auditor's opinion suggests inability to continue as a going concern
- reports of low employee morale
- passengers report clear deterioration in quality of service
- reports of management's lack of planning and direction
- articles questioning the airline's survival
- reductions in service due to loss of market share and sale of assets
- travel agents suggest alternate carriers to their clients

SUMMARY

Current research on corporate financial reporting practices and a review of recent airline financial reports suggest that airline companies have the ability and latitude to improve the financial appearance of the bottom-line earnings figure. Manipulation of bottom-line results can be conducted fully within the law, through the use of liberal applications of accounting principles. Perhaps more importantly, the bottom line can be significantly enhanced through management operational decisions, appropriately timed, which generate significant income or losses. Assessment of the true profit performance of an airline requires close scrutiny of the audited financial report and accompanying notes to become aware of liberal accounting methods, accounting changes, or specific transactions or events that alter the appearance of the company's true financial status.

PART 3

Current Issues and Problems in Air Transportation

Chapter 11

Economics

Courtesy of Boeing Commercial Aircraft; Boeing 757

DESCRIPTION

If one wants a definition of economics in air transportation, there is no shortage of supply. Broadly stated, it is a study of how airline managers choose, with or without the use of money, to employ aircraft and people to produce travel options over time and distance. The assets are distributed, now and in the future, among potential travelers. Air transport economics is complicated because it touches on many disciplines: social, financial, political and psychological, to mention a few. The airlines are capital-intensive, cash-flow driven, and competitively uncertain. The industry is strongly affected by the world's economy and by political events.

It is important to understand that the air transport industry has been a government-regulated public service industry worldwide. Various governments, both in the United States and abroad, exercise control over where and when the air carriers can fly and how much they can charge for their services. The essence of that type of regulation is the explicit substitution of competition with governmental controls as the principal device for assuring good performance. The stated goal of government regulation is to maximize economic efficiency while providing adequate and safe service to the public. This is the equivalent of minimizing the cost of providing any given level of service. The question of the balance of the "visible hand" of control and the "invisible hand" of competition is a central responsibility of the regulator. If extremely rigid controls are mandated, inefficient operations by the airlines could be the result. On the other hand, unrestricted and rampant competition could lead to excess capacity and heavy losses for carriers.

As we examine air transport economics, we will find that regulation of international carriers has continued, while deregulation in 1978 changed the economics of U.S. carriers. Thus, we will look at the history of air transport economics up to deregulation, then to the present, and finally examine factors of air transport economics as they exist in the industry.

HISTORY

The Jet Age that came into being in the 1960s also heralded the emergence of air transportation as a high-technology industry, and its economics rose threefold. America's position was awesome, for it dominated the production of jet aircraft (supply) and the majority of passengers on both domestic and international flights were Americans (demand). The U.S. aircraft manufacturers—Boeing, McDonnell Douglas, and Lockheed—had a comfortable lead time in heavy aircraft production and competed favorably with all comers. They were heavy in research and development, the travel industry was enjoying double-digit annual growth, and America was unchallenged. Signals of erosion in productivity and impending vulnerability were ignored.

By the end of the decade, the world economy and air traffic growth faltered. Not only was there a cyclical downturn but the long-term growth rate of air traffic demand slowed because the cost savings and added convenience derived from the introduction of jet aircraft had largely been achieved. Although the CAB reached a decision to raise fares in 1969, the board also informally began a route moratorium and ceased processing applications for new competitive routes.

In the 1970s, the industry entered the wide-bodied era, the single- to double-aisle airplanes. World economic and political events had an increasingly profound effect on air transport performance. Both the airlines and manufacturers missed the signals, and wide-bodied aircraft were placed in markets too small for the capacity. In 1973, there was a dramatic rise in fuel prices that began after the Yom Kippur War in the Middle East. Coupled with this was a background of accelerating inflation and monetary instability. Severe fuel shortages in 1974 forced the trimming of schedules and even the grounding of aircraft.

The world monetary system was further disrupted by fuel-related balance-of-payments problems. This complicated fares and rates negotiations. In 1975, there was expressed growing concern about the desirability of airline regulation. During the next three years, air traffic became erratic, inflation escalated, and airlines encountered difficult economic times. There was one bright spot in 1976 when the CAB relaxed requirements for charters. Trunks were allowed to match discounts and long-haul routes experienced growth. During the third quarter of 1977, traffic on the transcontinental markets increased as much as 61%. (Three years later, the three transcontinental carriers were joined by four more and all began to lose money.) The year 1978 was good for the industry. Despite an increasing rate of inflation throughout the economy, airline prices (in current dollars) declined for the first time since 1966, RPKs rose faster than any time in the previous 10 years, and industry operating profits jumped more than 50% above 1977. The explanation for this fortuitous event seemed to be the loosening of the regulatory grip on pricing.

DEREGULATION

The Deregulation Act of 1978 (ADA) was signed into law by President Jimmy Carter. It provided for:

1. **1981** Airlines determine their own domestic routes and schedules.
2. **1983** Airlines are free to set domestic fares; responsibility for mergers and interlocking relationships transferred to the Department of Justice (DOJ).
3. **1985** Subsidies for small community service transferred to Department of Transportation (DOT).
   Foreign air transportation rules handled by the DOT with consultation by Department of State.
   Foreign/domestic airline relations transferred to DOT and termination of the CAB.

The post-deregulatory competition was fierce. Carriers took off on reckless courses with marketing, capacity, routes, and pricing innovations. For example, Braniff Airways applied for more than 400 new routes, was awarded about 240 new segments, only to discover that it did not have the financial or human resources to serve the market start-up needs. It was not uncommon for carriers to price below total operating costs (TOC) or even direct operating costs (DOC) in order to capture market share, only to find that competition on the route matched them dollar for dollar *in a matter of hours.* The result was no market share capture, and total yields dropped dramatically. (See Figure 11.1 for sample analysis of operating costs.)

The number of airlines holding Certificates of Convenience and Necessity went from 37 in 1977 to 98 in 1982, which, added to commuter carriers (using aircraft seating 60 passengers or less, but not certified), raised the number of new entrants to well over 100 later in the decade.

The newly certified carriers were formed with used, often surplus, aircraft, low labor costs, and high productivity (no unions), which upset the industry's economic structure. Substantial economic turmoil resulted as the newcomers flew the highest-density routes, each trying to attract passengers with low fares. Markets were glutted with excess capacity. The majors' operating returns of $1.2 billion profit in 1978 shriveled to a loss of $600 million in 1982. Proponents of deregulation had predicted lower fares, greater frequencies, higher load factors, and better service—unachievable goals. These were essentially

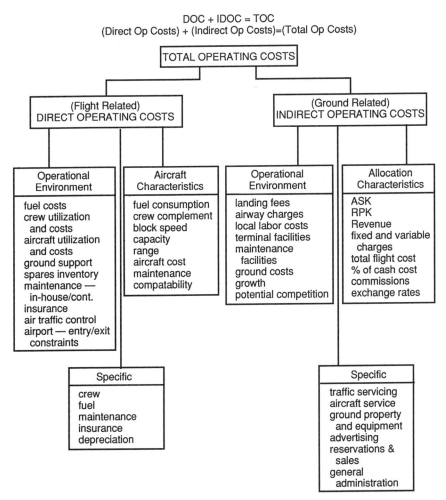

DOC + IDOC = TOC
(Direct Op Costs) + (Indirect Op Costs)=(Total Op Costs)

TOTAL OPERATING COSTS

(Flight Related)
DIRECT OPERATING COSTS

(Ground Related)
INDIRECT OPERATING COSTS

Operational Environment	Aircraft Characteristics	Operational Environment	Allocation Characteristics
fuel costs crew utilization and costs aircraft utilization and costs ground support spares inventory maintenance — in-house/cont. insurance air traffic control airport — entry/exit constraints	fuel consumption crew complement block speed capacity range aircraft cost maintenance compatability	landing fees airway charges local labor costs terminal facilities maintenance facilities ground costs growth potential competition	ASK RPK Revenue fixed and variable charges total flight cost % of cash cost commissions exchange rates

Specific	Specific
crew fuel maintenance insurance depreciation	traffic servicing aircraft service ground property and equipment advertising reservations & sales general administration

NOTE: The TOC for an aircraft within specific market pairs is variable, depending upon how and where the aircraft is scheduled. The DOC is dependent upon both the Operational Environment and the Aircraft Characteristics. The IOC is dependent upon both the Operational Environment and the Allocation Methodology. To understand the composition and breakdown of TOC, an analyst must thoroughly investigate the airline's procedures and objectives.

Figure 11.1   Analysis of Operating Costs

the benefits that advocates of regulation had voiced years before, and this called into question many government regulatory policies.

Additionally, deregulation in the United States spilled over into the international environment. All of the elements on which international economic air relations were based began to develop heavy stress

characteristics, both independently and as a result of interrelation-
ships. With the battle cry of "set the airlines free," Britain's Sir Freddie
Laker challenged the European internal network, hoping to bring
about changes as had his Laker Skytrain altered many of the rules on
long-haul international routes. Because deregulated objectives touted
high-load factors accompanied by reduced fares with improved service
and higher revenues, political leadership in a growing number of countries found themselves involved in a revolution to match these objectives. In less than 10 years, the number of carriers serving the Atlantic
routes rose from a handful to more than 50. The economics of comparative inelastic demand with a tenfold increase in capacity could only
result in horrendous losses for all but a few.

ECONOMIC CONSIDERATIONS

Having briefly examined the history of regulation and deregulation
from an economic perspective, we will now explore more closely some
of the factors of the air transport industry.

Given the current rates of inflation world-wide, cost-containment
measures are crucial to the entire industry and to those who regulate
and manage it. More efficient utilization of airline resources must be
achieved so as to keep costs, and ultimately fares, from reaching unacceptable levels.

In addition to cost minimization, a fare structure must be established that provides sufficient revenues to support an efficient and safe
air transport system. Airline fares should be adequate to cover all costs,
capital expenditures, and to earn the company a reasonable profit. The
definition of "reasonable" profit, and if the airlines are operating "efficiently," is at the discretion of the regulatory authority. A serious question is posed where authority is given to one body (the CAB) and
implementation to another (the airlines).

An optimal fare level is, in the words of a British transport report,
"the minimum economic price which can be contrived"[1] —"economic" containing the notion that an airline must earn enough to meet
its costs and to make an adequate profit. It is interesting to note that the
U.S. scheduled airlines achieved the CAB standard of a fair return on
investment[2] in only one year between 1960 and 1978. Either the CAB

---

[1]"British Air Transport in the Seventies," Report of the Committee of Inquiry into
Civil Air Transport (Chairman, Sir Ronald Edwards), HMSO, London, 1969, p. 9.

[2]The CAB's standard of a fair return on investment was 10.5% between 1961 and
1970 and 12% between 1971 and 1978.

established fares at an imprudent (politically motivated) level, or else the U.S. carriers consistently operated inefficiently (at least according to CAB standards).

Prior to the Jet Age, labor costs represented something less than 30% of an airline's DOC. From 1960 to 1978, labor costs crept up to account for 40% to 50%. In addition, productivity decreased. The impact on the air transport industry has been to make it unusually vulnerable to cost changes. (Fare increases to compensate for the increases took as long as four years of government hearings.) During times of high inflation rates, labor costs have risen faster than the inflation rate. There have also been frequent work stoppages or slowdowns, to which airlines are particularly sensitive, since it is not possible to warehouse empty seats and sell off inventory during strike periods. Slowdowns have been more devastating than stoppages, for during slowdowns, labor receives full pay and increases costs to an inordinate level with little productivity.

Other airline costs also increased dramatically, and none more so than fuel. Fuel costs were less than 20% of DOC prior to the Jet Age. From 1967 to 1977, fuel costs for U.S. airlines jumped over 210%, while the Consumer Price Index (CPI) increased only 71% during this period. The worldwide excess demand for crude oil pushed the prices upward rapidly, and fuel reached 40% of DOC. The magnitude of the problem for airlines is underscored by the fact that a one-cent-per-gallon increase in the price of fuel costs the world's airlines an excess of $1 billion annually. In the United States, each one-cent-per-gallon increase in the price of jet fuel results in well over $100 million in annual expenses for U.S. scheduled airlines.

Commissions paid to travel agents also changed in the 1969–1979 period, a growth of 220%. As travel agents sell as much as 70% of all tickets in the airline industry, they provide a distribution resource that carriers cannot duplicate. From 1960 to the fuel crisis period of 1973–1974, the world's airline business has seen its most rapid expansion; it doubled its traffic every five years. Travel agents are a critical part of the marketing of airline services and distribution of tickets.

Airport costs, including landing fees, rentals paid for terminal space, ground services, and airway surcharges, have risen rapidly in the Jet Age. From 1969 to 1979, landing fees increased almost 200% in the United States and were even higher internationally. At most airports, the revenue paid by carriers covers not only the majority of airport operating expenses but also the principal and interest payment on bonds outstanding for airport improvements. Other sources of airport revenues come from rentals paid by airport concessions, such as parking lots, restaurants, stores, and various shops. Thus, the airlines are compelled to support an ever-increasing debt burden for airport plant improvements, which are always occurring. One airline captain was heard to say, "I had a dream last night that I landed at an airport that had been completed." Costs related to ground-support services have increased since the Jet Age from about 8% of DOC to take a larger slice of at least double that percentage.

In the international aviation industry, in addition to the above, landing fees and ground charges can be completely arbitrary, frequently unrelated to the cost of the service. Some countries use airport charges as a competitive tool and/or to service other non-airport debts. It has been estimated that at Heathrow Airport (London) the landing fee for a fully loaded Boeing 747 at a prime time can be of the order of $10,000.

In Australia, there was a massive protest against increased landing fees. The fees were rescinded. The landing fee remained constant but the overall charges to carriers were raised beyond the proposed increased landing fee. It is interesting to note that this increase typically has little impact on the national carrier, which is owned or controlled by the same government agency that assesses and collects airport charges. Such payments might be looked upon as taking the money from one pocket and putting it into another.

Another economic concern of the airline industry is debt service costs. Airlines are short-term, cash-flow oriented, and voracious users of capital. They are highly leveraged, and technological advances have made it necessary for the airlines to re-equip their fleets about every eight years. Each of these cycles requires enormous outlays of capital. For example, the 12 U.S. trunk airlines alone spent about $6 billion from 1970 to 1972 to purchase new wide-bodied jets. A more recent purchase of 80 medium-sized Boeing 757 aircraft by American and United Airlines carried a price tag of more than $4 billion.

Airlines are experiencing difficulty raising the large amounts of capital they need. Their earnings records have been inconsistent over the last decade, with several years of low profits and heavy losses. Thus, few carriers can supply the capital from retained earnings. The dilution of equity renders raising money by offering new issues a difficult alternative. Consequently, airlines have had to rely more and more on debt financing. As the amount of airline debt has increased, and as earnings have remained inconsistent, the airlines' credit ratings have suffered. This has necessitated higher interest rates, borrowing for shorter periods of time, leasing aircraft under less attractive conditions, and other restrictive financing arrangements. The onerous debt service constraints under which airlines must function are setting up a vicious circle of debt, low earnings, more debt, continued low earnings, increased debt, etc.

The world economy is undergoing marked changes, and no industry is more sensitive to its economic cycles than air transportation. As mentioned previously, airline costs accelerate rapidly during inflationary periods. Because fare increases by government regulators or bilateral agreements normally lag behind airline cost increases, airline profits and reserves were further depressed during the last years of regulation.

The airlines entered the 1980s with seriously high debt/equity ratios. The erratic earnings record of the airline industry raises serious questions about the ability of the airlines to raise the capital they need in the 1990s. The Air Transport Association (ATA) has estimated that the U.S. airline industry will need approximately an average return on investment (ROI) of 13% to 15% to meet a $100 billion capital requirement in the 1990s. To support such an enormous financial need, the airlines must maintain a record of consistent profitability. It is estimated that about 50% of this investment will be for growth, the other 50% being used to replace the jet fleets purchased in the 1960s and early 1970s.

The economic dichotomy exists because technological advances were able to offset new equipment costs. In the 1960s, the airlines were able to finance the changeover from a propeller-driven fleet to a jet fleet with relative ease because of the tremendous productivity gains the jets brought. (Productivity is a combination of five factors: utilization, speed, passenger capacity, load factor, and service life.) The increased speed, capacity, and reliability of the jets revolutionized air travel. Direct operating costs in cents per available seat kilometers were reduced by almost 50% with the introduction of jet transports. That is to say, productivity gains in utilization, speed, passenger capacity, and service life almost doubled. The early, prosperous years of jet operations provided adequate internal cash flow to finance continued replacement of piston aircraft in addition to substantial industry growth.

The introduction of wide-bodied jets in the 1970s provided another giant productivity step of an additional 50%. But the cost of the wide-bodied equipment quickly outstripped the airlines' internally generated funds. The fuel crunch and labor gains increased DOC, and the recession a few years later impinged upon RPKs; both factors reduced the industry's ability to generate internal funds. After deregulation, low-cost airlines began to compete with high-cost carriers and created a competitive disequilibrium. Majors had to adjust or go out of business. Productivity gains were accomplished through increases in aircraft utilization. Seat pitch was shortened between rows and additional seats put into aircraft for gains in ASKs. Labor had to make concessions by agreeing to wage accommodations as well as flexible work rules.

During the first three or four years of deregulation, many factors caused the industry to falter. New entrants was one; the fuel price increase in 1979 was another. Commuter airlines grew rapidly and experienced a 43% jump in enplanements and about 75% in RPKs between 1978 and 1981. The air controller's strike of 1981 reduced available slots at major airports to about 75% of previous service. During this time it has been estimated that 40% of the fares were not related to distance but to competition. So, though load factors went up from the 50% to 60% range to 60% to 70%, total revenues (corrected for inflation) made only marginal gains.

In general, daily utilization increased by about 50% for all carriers as a whole. Utilization is a trap, however, because it purports to take economic advantage of IOC, which remains relatively fixed and allows the additional hours of flight at marginal costs. Yet, if these utilization increases capture small markets with low load factors, the utilization has not been optimized. The increase in daily utilization is only valid if the aircraft can be put to use in viable and profitable markets. Finally, aircraft speeds have remained relatively constant at about Mach 0.845.

Chapter 12

Cargo

Courtesy of British Aerospace PLC; British Aerospace ATP

DESCRIPTION

The air cargo industry typically functions in market pairs from medium to long distances, since handling and transfer costs render short distances unprofitable. Air cargo is thought of as freight, express, and mail. Freight accounts for more than 80% of all cargo in revenue ton kilometers (RTK). The balance is distributed between express and mail.

In the early years, little cargo was carried by air. In 1927, Pan American began operating the first overseas cargo service—a route that covered a 90-mile stretch between Key West, Florida, and Havana, Cuba. Even then, only occasionally were packages shipped, and then weighing not more than 5 pounds. In the same year, regularly scheduled Air Express Service was begun by the Railway Express Agency in the United States. However, the air cargo business really did not make big strides until after World War II. Yet, a short time ago, air cargo was limited to only a relatively small list of products. Although the airline industry currently carries only 1% of the world's total freight volume, that cargo represents 25% of the industry's total freight value.

STRUCTURE

Air carriers engaged in cargo are both passenger airlines and all-cargo lines. Passenger airlines transport cargo on passenger flights in the cabin and in the belly (lower lobe); carriers also often have dedicated cargo aircraft in their fleets and compete directly with all-cargo carriers.

It has been argued that passenger airlines, by and large, experience losses on their cargo services. There is extensive dispute over the profitability of carrying cargo on passenger aircraft. The debate stems from the allocation of costs by airlines between passenger and cargo services. Some passenger carriers take the position that the available cabin/belly space is essentially a no-cost facility of the passenger service. And, in fact, about 6% of passenger aircraft revenue is derived from lower-deck cargo.

Only 10 major carriers under section 401 of the Federal Aviation Act conducted scheduled all-cargo operations in 1978. By 1983, 109 carriers were certified to do so under section 418.

FREIGHT

Freight is the largest and fastest-growing component of air cargo. This growth has two sources: (1) the continued demand for air transport of time-sensitive or emergency shipments; and (2) the rapid replacement of surface freight by air freight as the selected transportation mode for nonemergency, large-volume shipments.

The basic characteristics of the market are:

1. Service (rather than price) sensitive. (In a recent survey conducted by a major air cargo carrier, quality of service was the main concern in 84% of customers using air freight service, whereas only 16% of the reasons given were related to cost.)
   –efficient handling of emergency needs
   –schedule reliability
   –pickup and delivery
   –prompt tracing
   –dependability
   –door-to-door time savings
   –cost
2. Value-to-weight ratio. (Electronic components have a high value-to-weight ratio, while iron ore has a low value-to-weight ratio.)
   –product markup (A high markup product can afford air; a marginal markup product might not.)
   –opportunity cost (physical, demand, perishability, product substitution, deadlines, breakdowns, shelf time, expansion to distant markets, flexibility, and product life cycle)
   –security (theft, damage, and loss)
3. Complementary mode of transportation (Occasionally a temporary support or when surface facilities are limited.)
   –surface unreliability
   –products change from price-sensitive to time-sensitive to serve peak demands (Surface serves long-time horizons; air serves short-time requirements.)

Air freight market share is less than 0.2% by weight but 16% of value.[1] Where these two factors are not constraints, surface is the preferred transport method. However, air cargo experts are finding ways to reduce cost, increase service, and reach wider markets.

Air freight rates vary by commodity, length of haul, volume and weight of shipment. The commodities, or types of freight, may have one or more of the following characteristics: valuable (women's wear), perishable (agricultural products), large-volume/high value (radar tubes),

---

[1]"Airline Planning and Marketing," U5010, January 1977, Boeing Commercial Airplane Company, p. 46.

large-volume/low value (TV sets), low volume/high value (computer chips), good density for aircraft (documents), fragile (some medical equipment), extremely valuable (diamond rings), or hazardous material (acids, flammable liquids). A number of methods have been developed to measure the potential in air eligibility. Generally, air transport is viable if freight is determined to be routine perishable, time sensitive, or surface transportation divergent.

The growth in air freight demand is, in part, attributable to the development of the concept of "total logistics costs." Door-to-door shipping by air has been the backbone of small-package courier services, and it has permitted reductions of stock inventory in diverse field warehouses. While in transit, air cargo spends much more time on the ground than in the air so that ground systems strongly impact on time and cost factors.

SMALL-PACKAGE EXPRESS

In the late 1960s, several airlines began to offer a premium-priced small-package service, whereby parcels handed over at the passenger counters were treated as unaccompanied baggage and loaded onto the next outgoing flight under a 50% refund guarantee. The shipper was required to make arrangements for pickup at the destination airport. The cargo traffic consisted mainly of cancelled checks, emergency repair parts, films, medicine, and legal documents.

An economics research paper at Yale provided Frederick Smith with several perceptions that the nation has become bewilderingly decentralized, that air freight shipments sent via passenger routes are not always traveling on ideal business routes, and that industry was willing to pay handsomely for speed, convenience, and service. Smith also recognized that commercial aircraft seldom fly at night, and, therefore, planes, airways, and ground facilities were in generous supply just when they could be used for overnight package delivery. Fred Smith used these ideas as the seed for Federal Express.

His strategy hinged on establishing a tightly controlled single hub (Memphis) and on owning his own jets (Falcons) so that he would neither be tied to commercial schedules nor limited to what could be tucked into the bellies of passengers planes. Within 10 years, Federal Express's revenues were over $100 million annually and almost a half-million shipments were made nightly. Federal Express spawned a new and viable segment of the cargo industry.

As evidenced by its record growth since 1973, the express air freight niche (high-priority parcels weighing under 70 pounds) tapped a need for fast, reliable transportation of small shipments. Despite the premium prices throughout this segment, shippers can economize in their selection of particular services.

AIR FREIGHT FORWARDERS

The airport-to-airport (line haul) of freight is only part of the air cargo system. The total operations also includes pickup and delivery, containerization, packaging, documentation, billing, and collection. Some shippers deal with an airline directly and deliver the shipment to the cargo terminal. Alternatively, the shipper may hire a general motor carrier to transport the shipment.

The domestic airlines in the United States, on a joint basis, formed an independent agency to provide their own fleet of pickup and delivery trucks for local service. The agency, Air Cargo, Inc. (ACI), divides its revenues and expenses proportionally among its owners. It assumes the role of an agent for the airlines, and it contracts with local truck operators for pickup, delivery, storage, and the issuance of shipment documents.

Others use freight forwarders who perform a role analogous to that of a travel agent in the passenger business. Under deregulation, the CAB declared that payments by the airlines to shippers, forwarders, and others preparing cargo for transport are no longer prohibited. Additionally, the air freight forwarder solicits freight, generally in small shipments. This individual can charge customers normal rates for the commodities and quantities they ship. The forwarder can also consolidate all of the cargo moving to the same destination and purchase bulk line rates (scale economies in pricing) and charge the shipper a lower rate than the shipper might get from the airline. The air freight forwarder breaks down the consolidated shipment at the destination and can perform local delivery. Because air freight rates include a wide variety of weight and volume discounts, forwarders are able to take advantage of multiple discount mechanisms to gain a competitive advantage over other modes of transportation.

The air freight forwarders are vital to the airlines, providing a large part of the air freight traffic. They deal directly with the question of who is responsible for pickup, delivery, consolidation, and documentation. Moreover, air freight is dominated by small shipments. Recent surveys have revealed that more than 70% of freight shipments weigh less than 100 pounds. The surveys have also indicated that most shipments consist of more than one package. By reducing the number of individual shipments tendered at the terminal and by increasing the use of consolidated shipments in their own containers, forwarders can reduce the paperwork and ground handling costs for the air carrier.

The government gave approval to the freight-forwarding function following World War II. The airlines were opposed to freight forwarders from the start on the grounds that: (1) use of forwarders would delay the movement of shipments by waiting for consolidation; (2) rate reductions on large-quantity shipments would be more difficult because the forwarders' income depended upon the spread between small and large shipments rates; and (3) freight forwarders represented an unhealthy potential concentration of buying power. Despite this opposition, freight forwarding grew rapidly. Less-than-planeload traffic continued to account for nearly all air freight business.

A few years ago a new trend developed to further complicate this segment of the industry. Surface carriers have obtained air freight forwarding rights. Also, not all air cargo firms can be described as forwarders. For example, the operations of many companies such as Airborne, DHL, and Federal Express include both the role of the air freight forwarder and the air carrier since the companies pick up the freight, fly it in a fleet of their own planes, and deliver it.

TECHNOLOGY

The air cargo industry has experienced drastic technological changes in many aspects of its operations. In addition to the profound impact of the jet engine and wide-bodied aircraft, there have been major developments in containerization, automated ground-handling facilities, data processing, and communications networks.

AIRCRAFT

In the past 25 years, the most publicized aspect of the growing air freight industry has been the development of larger and faster aircraft. The expansion of cargo traffic has been 3% to 5% greater each year than passenger traffic. The bellies of passenger aircraft provided an enormous amount of lift until the cargo market grew to such proportions that it would support aircraft dedicated to hauling freight. Experiments in the 1940s and 1950s were extensive. The Douglas DC-3 (unpressurized) was altered with a large cargo door, reinforced floor, and heavier landing gear to become an all-cargo C-47. The twin-engine Curtiss Commando (C-46) was a successful dedicated air freighter. The DC-4 (C-54) four-engine Skymaster was another unpressurized freighter with long-range capabilities. In response to freight lift requirements, almost every passenger aircraft could be converted to a freighter and was offered in that configuration.

When aircraft became pressurized, the conversion to all-freighter was more difficult and expensive, but almost every aircraft in the industry became a candidate. The conversion to an all-cargo model extended the life of a passenger aircraft for, after being replaced by more efficient new airplanes, it could be put through an airframe overhaul and be "zeroed" as a new "old" machine and serve a second generation of use.

Some aircraft were converted to extreme variations; the DC-4 had its nose enlarged to a huge bulb to accommodate automobiles, and was used to ferry cars across the English Channel. The Boeing Stratocruiser was converted to an oversized hull for carrying outsized cargo. It was called the "Guppy." The four-engine turboprop CL-44 had a swingtail. Several feet in front of the tail surfaces the entire frame swung wide open to accept cargo.

In the 1960s, most of the cargo was carried in the passenger lower lobes. Some all-cargo jets were built and proved to be successful. An attempt to offer the "Convertible" aircraft was particularly the strategy of the Boeing 727QC (for Quick Change). This trijet was programmed to carry passengers during the day and cargo at night. It was advertised as convertible in one to two hours from passenger to freight configuration. Experience showed the aircraft never to reach the levels of success anticipated, and for a host of reasons.

The conversion more often took longer than planned (often two to three times longer) and was costly. Moreover, in both passenger and cargo functions the 727QC was a poorer performer than the nonconvertible model. For example, the passenger configuration of the plane had to carry the extra weight of the cargo doors, the reinforced floor, and the heavier landing gear. The cargo configuration had to carry the galleys, toilets, life rafts, and baggage racks used in passenger mode. Nonetheless, four of the domestic trunk carriers (Braniff, Eastern, TWA, and United) adopted the convertible trijet. The concept was dropped later due to the operational limitations discussed here, depressed markets, and the impact of the large belly-load carrying capacities of the wide-bodied jets.

Another innovation was the "Combi," which would allow carriage of both passengers and cargo on the upper deck. This permitted the economies of scale in operating one large aircraft, which was flexible in the ratio of cargo to passengers, for seats could be added or subtracted. The "Combi" has found marginal success abroad as a special-purpose aircraft.

The costs of operating cargo aircraft are slightly different from those of passenger airplanes. Direct operating costs (DOC) of cargo planes include crew, fuel and oil, direct maintenance, maintenance burden, and depreciation. Indirect operating costs (IOC) include loading and unloading, landing fees, ground servicing, promotion, traffic agents and administrative expenses. In freight DOC, the crew does not include service personnel as on passenger flights, and depreciation is typically not as large as in passenger operations. Passenger markets are more sensitive to service and modern aircraft and other such competitive factors. Cargo markets generally assume a secondary position and, when possible, passenger aircraft are relegated to cargo tasks. The full depreciation has often been taken and, therefore, does not have the same impact as in the passenger environment. In IOC, loading and unloading is more mechanical oriented than people oriented and has a different cost-relationship within indirect operating costs.

The introduction of the wide-bodied jets into airline operations was a profound shock to the cargo side of the business. The first generation, narrow-bodied airplanes caused direct operating costs to decline on a per-ton/kilometer and per-kilometer basis over propeller-driven aircraft. Then the wide-bodied jets lowered DOC per-ton/kilometer due to economies of scale, though the DOC per-kilometer was raised by 75% to 100% over the narrow-bodied jets.

Consider that the passenger model 747 had 5,550 ft^3 of usable lower-lobe space, compared with 7,610 ft^3 in an all-freighter 707. Boeing literature emphasized that, while the 707 all-freighter would break even at 50% load factor, the 747 all-freighter would do so at 33%. What was not made evident was that, since the 747 was so much larger than the 707, 33% load factor for the former craft would equal double the load of the latter at 50%. (In fact, one airline advertised that each of its 747 freighters could carry the equivalent of five boxcars of freight at 600 mph.)

In addition to the enormous capacity of jumbo jets, one must consider that the average speed of these aircraft is 30 times greater than that of a freight train and 40 times faster than a trailer truck. Yet, most air freighters are cargo versions of passenger aircraft with concomitant diseconomies.

Costs

The total operating costs (TOC) for an airplane over a specified distance are variable, depending upon the method of cost allocation and on how and where the aircraft is used. Direct operating costs are dependent upon the operational environment as well as the airplane characteristics. In terms of trip length, total DOC will increase with distance but be reduced per-kilometer or per-hour basis. Should an aircraft fly 10 one-hour cargo flights in a day, the landings and takeoffs increase the kilometer and hourly cost far beyond those when making one 10-hour flight. Indirect operating costs are dependent on the operational environment, but also upon the allocation method used, in terms of specified distances. It can be said that indirect operating costs are relatively insensitive to trip length and time.

Aircraft capacity involves both weight and volume, and it has a profound effect on costs. First-generation jet cargoliners "cubed out" (used all of the available space) before they "weighted out" (reached maximum allowable weight). The traffic carried by narrow-bodied jets has a lower density than the original design. For example, the design density of the 707 was 12.9 lb/ft^3, where the density of commodities loaded varied between 5.3 to 20.0 lb/ft^3, depending on the shipper. The average density was 8.6 lb/ft^3; therefore, a 707's effective cost per available ton-mile on a basis of "cube" would be 1.5 times that based on "weight" as long as the density of cargo remained at 8.6 pounds.

The manufacturers of the second-generation wide-bodied jets responded by designing the aircraft (the B-747) to 10.3 lb/ft^3 and the DC-8-60 to 9.9 lb/ft^3. Therefore, the DOC on a ton-kilometer basis was reduced significantly.

Figure 12.1 Glossary of Commonly Used Air Cargo Terms

Air waybill The document that is written proof of the contract between the airline and the shipper for the transport of the consignment.

Consignor The person or firm whose name appears on the *air waybill* as the party contracting with the transporter for carriage.

Consignee The person or firm whose name appears on the *air waybill* as the party to whom the shipment should be delivered.

Forwarder The person or firm expediting the goods to the *consignee* on behalf of the *consignor.*

Agent This is a term used for a *forwarder* when representing a particular person or company. In the case of IATA, a cargo *agent* is one recognized and approved by IATA. An *agent* may also be appointed by a specific carrier or group of carriers.

Intermediary When a *forwarder* represents many parties, this person may also be known as an *intermediary.*

Integrator A person or firm who has integrated, under a single administrative control, those functions normally undertaken by separate entities such as the *shipper, trucker, forwarder, agent, broker,* etc.

Customs Broker A licensed professional who processes *customs clearance* procedures on behalf of the *consignee.*

Express Cargo This is a loose definition. It refers to particularly time-sensitive air shipments that require expeditious customs clearances. IATA has defined it as "traffic requiring reliable, time-measured transport, normally on a door-to-door basis, using simple documentation, for an inclusive price with one carrier exercising integrated informational control."

Door-to-door An integrated service for which one party takes complete responsibility for the service. The customer pays a single tariff. It is typically not a price-sensitive market.

Courier A person carrying the time-sensitive or secret documents or small parcels through the passenger baggage channels. This avoids more time-consuming *customs* procedures for air cargo. It is also *door-to-door.*

Hard freight Large air freight whose physical measurements place it outside the traffic normally shipped by *integrators* or *express operators. Hard freight* requires the use of mechanical handling equipment, and it is considered the domain of traditional air carriers or dedicated air freight carriers.

SUMMARY

Of all the tasks that air freight management must face, identification of the market is the most important. Unless management is able to quantify and qualify the market needs in routes and service, profit and success are unlikely. Integrating the cargo function into the airline's overall objectives is highly subjective, but crucial for air freight management. If management does not target a market that can be served by the passenger flights, it can be a financial disaster, leading upper management to believe, erroneously, that cargo markets do not exist.

Management must not pursue cargo customers if it cannot match competition, for it can provide inaccurate signals. Numerous times, cargo markets have been overlooked and exploited by others. The small-package, overnight segment of the industry lay fallow for years until Fred Smith properly articulated it with the Federal Express service. The "Pregnant Guppy" conversion, which was the name of the propeller-driven Boeing Stratocruiser with its fuselage expanded, found a second economic life (15 years) carrying outsized cargo such as the third stage of the Apollo rocket, fuselages of new aircraft, and other cost-insensitive freight.

The selection of new cargo aircraft typically rests on

1. passenger demand (80%) for passenger air carriers;
2. both passenger demand (60%) and cargo capacity (40%), in the case of the combi-aircraft;
3. cargo requirements (100%), in an all-freighter airline.

A close match between capacity and demand can only be achieved if the correct aircraft is chosen; thus, the extent to which air freight management can influence acquiring an airplane type, which is responsive to its needs, may describe profit or loss in air cargo.

Inextricably linked with the cargo service features is pricing. Air freight tariffs, reported as yield per revenue ton kilometer (RTK), reflect a range of demand/cost factors. On the demand side, yield is affected by the nature of the commodity, type of aircraft, and containerization.

In view of the many variables affecting the rates of individual shipments, some might consider it appropriate to utilize a carrier's average yield per RTK as a proxy for its pricing policy. Obviously, the average yield is affected directly by the traffic carrier.

How an airline perceives its cargo service greatly impacts on its performance. If air freight is looked upon as a secondary source of income, it probably will not do well. This is not because demand does not exist, but for the reason that passenger service emphasis does not meet cargo necessities. Viewed in its true perspective, cargo can make an airline a profitable operator if the proper match is made. At only an incremental cost, it can occupy leftover passenger capacity and generate unexpected revenue.

The air freight segment of the air transportation industry is a straightforward part of the business. If the past holds firm, the industry can look forward to above-average growth by comparison with the rest of the airline business.

It is also true that there is a close correlation between key economic trends and air cargo growth. In the matter of airplane capacity, it is a matter of both weight and volume; each is extremely important. Air carriers that have been able to determine cargo requirements and make that a part of fleet buy decisions have optimized a significant revenue resource. Cargo capacity can be represented by three segments: lower lobes of passenger aircraft, upper decks of combi-aircraft, and dedicated freighters.

There are major considerations in developing criteria for measuring cargo demand: physical characteristics of cargo, service levels relative to flight frequency, time-of-day availability, and competition. The cargo system of a carrier is affected by the terminal as well as associated ground systems and support activities. Creative cargo marketing is a key consideration as containerization, consolidation, and small-package innovation have shown. Cargo handling is taking on more and more of passenger imperatives such as quality of handling, schedule reliability, ground support, elapsed time, market pairs, cost, and in-flight services. Cargo revenue at an incremental cost is an important consideration and is becoming a recognized segment of an overall strategy in planning and fleet selection.

Chapter 13

Trends

Courtesy of Airbus Industrie of North America; Airbus A-310-300

DESCRIPTION

An old Chinese proverb holds that, "To prophesy is extremely difficult, especially with regard to the future." The air transportation industry has moved from the Propeller Age through the Jet Age to the Deregulation Age, and each is vastly different.

The proponents of airline deregulation maintained that deregulation would produce a more efficient industry. The events of the past support this prediction, although the transformation has been extremely difficult and far from complete. Air carriers will have to continue to restructure their routes and reorient pricing policies (or the government will regulate them) leading to an industry more responsive to consumer demand. Productivity will have to continue to improve as efficient carriers grow, new entrants expand, inefficient carriers contract, and as the industry sheds costly practices that were nurtured during 40 years of protective regulation. The distortion imposed by regulation is now almost gone, and carriers will have to continue to make dramatic changes to survive.

EQUIPMENT

A large percentage of current equipment is relatively old and is bound to be replaced in the next decade. Because the development cycle of a new-generation airliner is lengthy and will cost a manufacturer an investment of at least the firm's total net worth, it can, with great confidence, be stated that the equipment introduced by the major manufacturers, the 757, 767, A-320, and MD-11, will be in service until the end of the present century. The new aircraft now being introduced by manufacturers contain several new technologies that were unavailable to the first-generation jet airlines planners; these consist of high bypass engines and electronic controls, flight management systems, noise reduction technology, airfoil and wing improvements, digital electronics, sophisticated structures and materials, and manufacturing productivity.

The transition period to the new-generation aircraft will probably occur in the early 1990s with the aircraft entering the transportation system in the mid-1990s. New-generation aircraft will be more fuel-efficient and quieter, since improvement in engine technology coupled with quantum leaps in airfoil and wing design will reduce fuel consumption by more than 30%.

Introduction of new materials such as composites will result in significant weight savings, enhancing the capability to carry additional passengers. Most commercial jets flying today feature "analog systems" that are both cumbersome and expensive to maintain. The Boeing 757 and the 767 as well as the Douglas DC-9-80 have made enormous improvements with sophisticated flight support and information systems. The new-generation aircraft contain a "star trek" automated and digital command system in the cockpit. This results in a less complicated human factor control of the aircraft and reduces flight decisions made by personnel. Moreover, there are significantly reduced maintenance costs (by 50% to 75%) and increased acquisition costs (50%). Furthermore, these digital systems will improve aircraft performance, increase reliability, and permit greater utilization. Electronic flight systems add to the productivity of the industry because of more accurate holding, precise approaches, and runway utilization, as well as enhanced ability to maintain separation criteria.

It will not be until the next century that there will be an introduction of alternate fuels such as liquid hydrogen. Liquid-hydrogen-powered aircraft appear to have numerous advantages but the state of the art must advance far beyond what it is. It would appear reasonable that liquid hydrogen will be applicable to large aircraft only. Furthermore, the use of this type of fuel will require re-equipping the world's major airports to handle liquid hydrogen, a task requiring a fantastic investment.

In the even more distant future one can anticipate the appearance of nuclear-powered aircraft, but this option will be beyond our planning horizon. The speed of sound, which is reached at around 640 mph at high altitudes, seems to set a real barrier to further speed improvements in conventional jet aircraft. The breakthrough was to have been the supersonic jet (SST). The United States gave up on the SST after substantial research. The British-French consortia built the supersonic Concorde, and the USSR produced the TU-144, to capture the lead in supersonic flight.

Both have proven to be unfortunate decisions in a commercial sense. The Concorde cannot generate enough revenue to pay fully allocated direct operating costs though it is a technical masterpiece. The Russian supersonic jet experienced accidents that cast some doubt as to its technical excellence, let alone its economic viability. Therefore, it is unlikely that aircraft speed will increase significantly until after the year 2000.

Besides the developments described above, there is another trend that has developed in addition to derivative models of aircraft. Of course there is the family of aircraft of the A-300, Boeing 757/767, and MD-11, as well as modifications to all of the aircraft presently flying the world's air transportation routes. Some re-design of existing aircraft has been taking place with more fuel-efficient and quieter powerplants to extend the lives of airliners. This trend is enhanced by the dramatic escalation of the cost of new aircraft, the inability to sell older airplanes, and the availability of spares support, all of critical importance to a cash-intensive industry.

Since many of the jets flying today are structurally capable of flying for a long period, it is cost-effective to reduce operating costs with fuel-efficient engines or to modify engines to lengthen the useful life for 10 or more years with a relatively low investment.

Micro-changes are taking place and will continue to be studied and evaluated. Airliners are being modified with wide-bodied interiors and lighter-weight seats estimated to recover costs in about two years. TWA embarked on a weight-reduction program that company officials thought would save about $2 million annually. Eastern Airlines found that removing the paint from a 727 aircraft saved the weight equal to that of an additional passenger and baggage, increased cruising speed at the same fuel consumption, and added one airplane to the working fleet as one aircraft had been in the paint shop on a daily basis. The list of micro-changes is almost endless in the industry as aircraft are modified to be more efficient.

To transcend all of the slow pace of growth of development, most breakthroughs are made during wars. Should there be another conflict, the military establishment will undoubtedly make exponential steps into hypersonic craft capable of Mach 5 (4,000 mph) or more and cause commercial aviation to jump by a generation.

TRAFFIC

Reviewing the trends of the world passenger traffic shows a declining growth rate as the industry matures. From 1929 to 1946, growth rates were substantial: almost always greater than 20% and averaging 30% for that 17-year period. The following 23 years saw the average growth rate cut in half to 14.5%. In the years between 1970 and 1975, the growth rate dropped to 9% on the average. Between 1976 and 1980 the compound growth rate was 8.54% for all services. The years between 1981 and 1985 saw growth at about 6.19%, while the balance of the century should see an annual growth rate slightly over 5%. In fact, such a growth rate is to be expected based on general historical trends. Long-term traffic growth is linked to the increase in GNP and, since the predicted rate of economic growth in the 1990s for the United States and other industrialized nations is about 4% to 5%, airline traffic estimates of about 5% for this decade will probably hold true, and many airline economists forecast the 5% growth rate through this century.

CAPITAL INVESTMENT

A comparison of capital needs of the industry during the last 20 years and the next 20 years clearly demonstrates a possible limit on growth. During the 1960s, the industry invested about $9 billion in equipment. This investment rose to $15 billion in the 1970s, and during the 1980s the capital needs of the airline industry have been about $60 billion in the United States and $110 billion worldwide. The dramatic increase in capital investment during the 1980s stems from increased traffic and accelerated replacement of obsolescent equipment by new and efficient aircraft, which, because of high inflation rates, are becoming more costly.

Having stated the capital needs of the industry, we turn our attention to the airline industry's ability to meet those needs. The early prosperous years of jet operation provided internal cash flow adequate to finance both the continuing replacement of piston aircraft and substantial industry expansion.

During that period, airlines were able to acquire additional long-term debt and equity capital under reasonable terms. Beginning in 1967, however, total capital needs began to exceed internally generated funds. By 1970, this gap had widened significantly, and since then capital investment has consistently outweighed the industry's internally generated funds, which has caused a marked deterioration of equity. Much of this discrepancy has been temporarily overcome by unusual long-term financing arrangements for jet equipment and aircraft leasing.

McDonnell Douglas has offered a no-down payment and as-you-go easy terms on its DC-9-80 aircraft, which has proven a boon to airlines in need of new equipment and short of cash. Eastern did not need the 200+ passenger capacity of the airbus A-300B, so the French struck a deal of selling the aircraft according to the required capacity instead of the actual seating.

For a period of time, the lease payments were structured as though the excess capacity did not exist. It was truly creative leasing by Airbus's George Warde. Today, however, outside sources of capital and government participation are limited; the industry's ability to borrow or lease is affected by its continued low and unstable financial earnings record.

In the 1990s both the U.S. domestic and world aviation industry will require at least a 350% increase of capital funds over that needed during the 1980s. Opinions differ about the availability of that amount, and it is clear that the airlines will have to compete for funds in a limited capital market. However, based on past earnings and the tarnished perception of commercial aviation, airlines will be hard-pressed to compete effectively with U.S. manufacturing enterprises for capital funds.

The U.S. manufacturing community profit margin has averaged about 5% during the past 10 years, while the airline industry has averaged almost zero. The airlines' dismal profit margins are another vivid illustration of the need to restore the industry's financial health before the severe capital demands begin in the next few years. Pressures and constraints impacting on airline growth are shown in Figure 13.1.

DEVELOPING SEGMENTS

The market for scheduled air travel is in its mature state and is linked to economic growth (see previous discussion on Traffic). Unless the airlines can reduce their costs, new entrants, not so burdened, will continue to capture traffic. The airlines' inability to rationalize their pricing as a cost-plus strategy instead of a competitive strategy has made them economically unsound. It may force the government to re-enter regulation with a minimum cost pricing. There has been considerable movement by unions to keep union jobs, and the problem of the industry has been the high cost of operations.

Those airlines able to adapt and meet the low-cost operators will survive, but it clearly describes the trend of air transportation toward mass transit, through lower costs and better management.

Charter groups will allow for economies of scale for many passengers. As about 70% to 80% of all travel is sold by travel operators, their focus is on selling one group rather than one person. Charters and tours are being designed for specialized groups: scuba divers, hikers, believers in the occult, and others. They are being organized as all-inclusive tours that permit the traveler to know what the total cost will be. There are enormous economies for the tourists and for the tour operator. More tours and charters for business corporations (which can be written off) will be sold. This trend is especially strong in Europe and now in the United States.

SUMMARY

The prospective market is not growing robustly. The reasons are because, first, there has been a slow traffic growth since 1980 (in 1982 world traffic was up only 6.3% from 1979) and as these rates are compounded annually, forecasts start from a smaller base than expected;

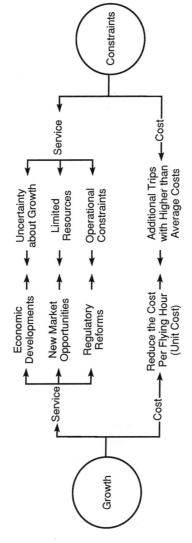

Figure 13.1 Pressures and Constraints on Airline Growth

second, this slow traffic growth rate influences projections and tends to reduce the outlook for traffic demand. A general estimate of traffic growth until the end of the century is about 5.7%, down from the previous 6.32% rate. As more and more of the world becomes deregulated and airlines have to replace their obsolescent aircraft with new equipment, fares will rise and this should have an effect on the growth rate.

Some of the crucial variables and conditions that will shape the course of air transportation for the next decade are:

1. Market growth rates will tend to be linear and inelastic, unstimulated by market variations, except in isolated cases.
2. As markets expand, specialization will continue to be a strategy engaged in more and more.
3. For the near future, aircraft builders and airlines will concentrate on extracting the maximum benefit possible from technology existing in current aircraft (A-320, MD-11, 757 and 767).
4. Retirement of aircraft will be about 5% of each year's fleet.
5. On the other hand, the size of the fleet actually becomes a product of the judgments of traffic, aircraft size, utilization, speed, load factors, and demand. The growth rate of fleets will probably be 3.8% each year.
6. Utilization could rise about 3% annually by better scheduling and usage.
7. No totally new technical advances will emerge until after 1995. This means that the next major shift into all-new generation aircraft will occur in the late 1990s.
8. Modification of present aircraft will continue at an increased rate.

As the development of air transportation continues, terminals will have to be expanded—and entirely new airports will have to be built. They will be larger than current ones and farther from city centers (such as Narita Airport in Japan).

Low-cost service, offered on a first come, first served basis is breeding a new awareness in air travelers. There is a continuing and apparent shift from luxury service into a commonly used form of mass transportation. Although gradual, the evolution of air travel into mass transit should be completed by the year 2000.

Chapter 14

Entrepreneurship

Courtesy of Dornier Aviation (North America) Inc.; Dornier 328

DESCRIPTION

There have been more than 100 new airlines formed since deregulation in 1978, far more experiments, and a large number of innovations among mature carriers. The basis for this transformation has been the emergence of entrepreneurial and innovative strategies practiced in the air transportation industry as never before.

These deviations from the *regulated* method of conducting the airline business demand an analysis. Before 1978, Competitive Variables were challenged in a political, constrained, and unpredictable fashion because of regulation. With government subsidy, control, and intervention, it was difficult to assess the effects of a Competitive Variable. In the deregulated environment, however, competitive decision making has exploded into a free-wheeling, multiple-option process with results mostly unsullied by subsidy or government controls.

Entrepreneurs and innovators have created new ways of competing in the industry. Start-up carriers have intruded on the profitable and high-density routes, and current airlines are experimenting with innovative actions to meet the increased competition that has eroded market shares and yields. As a result, entrepreneurship can be discussed as both the start-up of a new enterprise and the extension of an existing business.

It is well known that entrepreneurship, invention, and innovation are closely related: Entrepreneurship is the total start-up process; invention is the idea that "gels" and "flips the switch" somewhere in the mind of the entrepreneur; and innovation is the unit of change and the coupling process among idea, entrepreneur, resources, technology, and marketplace.

The environments of airline innovation are summarized in Figure 14.1.

THE 80/20 RULE

This differentiating will apply to both start-up ventures and mature airlines that are examining creative strategies that challenge the traditional way of doing business. The methods used to exploit Competitive Variables are common to both start-up and to operating airlines. To view statements made here as absolutes would be a gross error because this is one perspective of the volatile lifestyle of the industry and the various tactics being used to identify and exploit market niches. One should rather apply the *Pareto principle* to most declarations, which is the belief that the activity will fall into an 80/20 relationship.

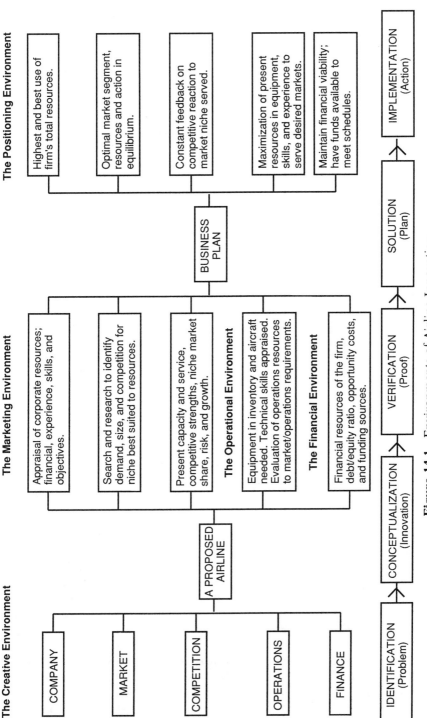

Figure 14.1 Environments of Airline Innovation

For example, it is said that executives spend 80% of their time on 20% of the problems; 20% of an airline's routes generate 80% of its total revenue; and a route monopoly might be considered to be an 80% or greater market share.

Entrepreneurship is dynamic, inconsistent, unpredictable, challenging, and only occasionally rational. Invention is creative action. Innovation is a process for which there is no set mold because what works for one carrier might not provide success to another.

In *Strategy-Making in Three Modes*, Henry Mintzberg wrote that the "entrepreneur was first discussed as that individual who founded enterprises."

In the words of Joseph Schumpeter, the late Harvard economist, "The entrepreneur finds capital which he brings together to form new combinations." More simply put, the entrepreneur is the mover-and-shaker. In *The Organization Makers* by Orvis S. Collins and David G. Moore, it was noted that "entrepreneurs are forceful men. They do not fit their desires and ambitions into a time schedule for movement up the ladder." That is, entrepreneurs often have no sense of work hours, but do have a sense of purpose. Further, these men and women know how to identify and exploit competitive opportunities in nontraditional ways and still produce successes.

AIRLINE ENTREPRENEURS

Airline entrepreneurs are a diverse breed: Stephen Quinto at Northeastern International Airways; Irv Tague at Midway and Connectair; Herb Kelleher at Southwest Airlines; Don Burr at People Express; Fred Smith at Federal Express; Freddie Laker at Laker Airways; Eli Timmoner at Air Florida; John Emery at Emery Air Freight; and Ed Beauvais at America West Airlines are just a few. In the early 1930s, the world was introduced to a cadre of frontier entrepreneurs: Pat Patterson of United; Bob Six of Continental; Eddie Rickenbacker of Eastern; C.R. Smith at American; Howard Hughes of TWA; and Juan Trippe of Pan Am. All of these men were risk-takers. They all had the vision and the ability to make adjustments when they were wrong. They properly identified a market need. They were lucky, and their timing was fortuitous.

Looking at the entrepreneur, one sees that they are a special breed of person. John Jakes might have been writing about airline entrepreneurs in his book *North and South* when he observed that "the race is to the driven, not the swift." The entrepreneur is a pusher, a plodder; his rate of speed is not as significant as constant forward mobility.

It appears, moreover, that entrepreneurial interests are more the accumulation of practical readiness than of individual psychology or personality. This is not to say anyone can play the game. It does suggest that people can become entrepreneurs at any time in their life and the inclination to take this leap is increased by their awareness and recognition of this as a career alternative that can be successful.

It is also important to note that management skills are not the same for every carrier's situation and may vary with the airline's position on its growth curve. Frequently, managers who possess "the right stuff" at certain levels are disasters at others. Entrepreneurs are often unable to cope with an organization that has emerged from a dynamic growth stage and is stabilizing. Howard Head found this to be true at his company Head Skis and then again at Head Tennis. He was excellent at designing, inventing, and innovating but was a notoriously poor manager. When each company matured, Head proved that entrepreneurs do not necessarily fit into all stages of a corporation's life. He was compelled to leave.

One of the tenets that has been recognized by the Taylor school of management scientists is that management techniques are often transferable from industry to industry (Pareto). Dr. Ted Leavitt at Harvard observed that the three most important elements in the success of a firm are "Management, management, management." That might explain why Al Casey was brought into American Airlines from the Parent Co. with no airline experience and turned the company around neatly in a matter of one year. At the age of 59 and with no airline experience, Ed Carlsen took over the leadership of United Airlines. The "Friendly Skies" soon became the largest single airline in the Western world. The list could go on and on, supporting the notion that managing is a special transferable ability from industry to industry in most cases. That is not to say that grooming management inside is not viable. Several airlines insist on bringing managers up through the ranks and have been quite successful at doing so. Delta is one example, Northwest another.

MANAGEMENT LEVERAGE

There is often a question raised as to the relationship of the various functions of a new enterprise. A suggested breakdown is that management represent 60%, market demand 20%, operations 15%, and finance 5%. Clearly, this is subjective, but the spread in percentages is intended to show the relative importance of various components in entrepreneurship and is consistent with other businesses.

What we should examine are (1) the characteristics of entrepreneurs and innovators as well as (2) the process of change and its behavior. Tools for effecting innovation and change will also be explored.

ENTREPRENEURIAL CHARACTERISTICS

Entrepreneurship is the total creative discipline. It is a departure from what exists and is a new form. It is typically the exploitation of a crack or niche in the market. It is usually short term in nature but, depending on the behavior and scope of the creation, can be long term.

It is a new airline start-up. (Herb Kelleher and Rollin King should have known that Southwest Airlines could not work, but it did.) It is a small-package express business. (Fred Smith should have known that there was not a market for small packages, but there was.) It is a new market for an established carrier, or new uses for new technology like the Concorde.

PROCESS OF CHANGE

Of all the change mechanisms, start-ups have a higher degree of uncertainty and more dynamic, if not chaotic, performances. A recent study has shown that more than 60% of innovations are incremental changes of existing technology, not dramatic, earth-shattering revolutions.

America West Airlines uses existing Boeing 737 aircraft on profitable routes, offering discount service to passengers in the western United States. By carefully niching themselves and growing within that niche, they have developed into a serious competitor for the larger domestic airlines.

MOTIVATION

A study of several hundred successful start-ups in all industries has shown that the most successful entrepreneurs have college degrees with some additional education. There are, however, many with no college education, many with graduate degrees, and even Ph.D. entrepreneurs are not uncommon.

Venture capitalist Hal Nissely has commented, "Some successful airline entrepreneurs had a big advantage, they never took business school courses so they did not know that some things would not work."

Degrees are important credentials for entrepreneurs wishing to secure venture capital and are formidable tools in the right hands.

Education is not, however, a guaranteed ticket to success any more than the right hammer will make a good carpenter.

Successful entrepreneurs were not motivated by money. That is what Tom Peters alluded to as the intense desire to accomplish something. *In Search of Excellence* describes this as the need to give birth to an idea, develop it, and bring it full circle to fruition.

Edward Schon at MIT wrote that "Champions of new invention display persistence and courage of heroic quality." This is the root of the Japanese philosophy of participatory management, where everyone is involved, the sharing and nurturing. It is more than dollars or yen; it is togetherness, achievement, recognition, and persistence. It is making a new idea a success that keeps them going.

It is astonishing to hear how often entrepreneurs referred to their failures before successes. Adam Osborne of the ill-fated Osborne portable computer said, "You've got to fail some of the time or you aren't trying hard enough." Entrepreneurs delighted in recounting their mistakes. It almost adds up to the fact that the successful entrepreneur can be measured by the number of mistakes made (Pareto). That may be an unsettling observation, but when viewed in its totality it becomes a crucial piece of information. They are risk-takers, not always right, but willing to take the chance. The underlying message is that entrepreneurs are not afraid to be wrong or to admit it. They march on, in the face of failure. They are doers, flexible and quick to feint and take corrective action when necessary.

Those who do nothing and take no chances can boast of never having made a mistake. They are colloidally suspended, merely treading water. They do not go down, but then they cannot go up either. In the "Peanuts" cartoon, Peppermint Patty says to Marcie, "I like Saturdays because I can't get any D's." The entrepreneur would point out that Patty can't get any A's either. She is not a Saturday thinker.

THE UNREASONABLE PERSON

Whether an airline is in a start-up position or is presently operating, whether it is a major or a regional, the willingness to take a risk, to move out of a protected position, is critical for entrepreneurship.

George Bernard Shaw was right when he said that "The reasonable man adapts himself to the world, the unreasonable man attempts to adapt the world to himself. Therefore, all progress depends upon the unreasonable man."

Another observation is that hours were not a significant factor in the workday. There was no correlation between success and hours on the job. Though most entrepreneurs devoted long hours and weekends, there were many who elected to work a normal day or even a shortened workday. The management of time was crucial.

Irv Tague took over Hughes Air West for Howard Hughes when the carrier was a strange mix of problems. Loaded with golden parachutes and disillusioned unions from the two previous mergers, the airline was losing more than $15 million a year. When Tague took over the reins, Hughes Air West had cash for only 15 days of operations and the Hughes Summa Corporation would not provide any additional supporting funds. Tague had the airline profitable in little over a year. What kind of hours did he work? He arrived at the office at about 10:00 A.M. and left between 2:00 and 3:00 P.M. Of course, there were exceptions, but this type of short day was his basic schedule.

Tague selected four outstanding officers, gave them full authority, responsibility, and accountability. He managed in a style that was incredibly successful. His sense of the quality of time surpasses most other people. When Tague left Hughes Air West his successor worked very long hours. Within two years the carrier was sustaining heavy losses and was sold to Republic Airlines. Although a simplified account, this is essentially what happened at Hughes Air West. No particular management style is being advocated here. The story serves to highlight the importance of productive, quality time over mere hours in the office.

Luck and timing are two necessary random variables in entrepreneurship. Midway Airlines reached the market at an ideal time although timing was not a part of their strategy. Luck found Southwest Airlines. At a time when the carrier was within reach of bankruptcy, it made an enemy of Braniff. Southwest was David to Braniff's Goliath. Over and over again Braniff announced that it was going to put Southwest out of business; people finally took notice. Southwest gained credibility and notoriety, and subsequently became quite profitable.

HOW IS IT DONE?

Of course, entrepreneurship is not at all a matter of daring. There is a nuts-and-bolts side to any new airline. First, resources must be appraised to determine whether there is a good fit between the market opportunity that has been identified and the ability of the airline to take advantage of it. This appraisal is pivotal to the success of entrepreneurship. Only by knowing what skills, finances, experience, and motivation that an airline and employees possess can management properly analyze opportunities present.

Second, the entrepreneur must select a management team that has the right personality for the airline or market segment opportunity—in short, all highly motivated achievers. Without the appropriate management team, small in number and of one mind, market penetration will be an exercise in futility.

The third characteristic of successful entrepreneurship is a proper assessment of market demand. John Newhouse, in *The Sporty Game*, referred to market need when he wrote of it as "a hole in the market." Unless the entrepreneur identifies and evaluates market demand, there is little or no chance for success, even if the right management team is in place and adequate financial resources are available. But even that is not adequate; the entrepreneur must also weigh the market demand against one's own resources to see if they balance.

What good is "a hole in the market" if the entrepreneur lacks the ability to fill it? A commuter airline cannot respond to an obvious opportunity in the transcontinental market; a major does not have the resources appropriate to a commuter niche; and a carrier dedicated to cargo service would not try to dig out a passenger hole in the market.

For example, Air Florida was initially a regional/commuter carrier, and faded into a bankrupt oblivion having attempted long-haul expansion. United Airlines abandoned its regional/commuter services, and World Airlines could not become a success in the scheduled passenger service regime.

The fourth requirement for success relates to the third. Once the entrepreneur identifies the market, the operations must be scaled to meet its requirements optimally. Supply must match demand in terms of capacity, frequency, range, and utilization. For example, a carrier may have the correct equipment to serve its selected market niche, but not the time due to high utilization. The result is that the carrier does not, in fact, have the equipment available.

The fifth necessity applies to all successful ventures. The entrepreneur must draw up a strategic plan with a definite strategic objective. In *Megatrends* John Naisbitt wrote that "Strategic planning is worthless—unless there is first a strategic vision." He also pointed out that not only must the purpose be right, it must also be shared. A shared strategic plan makes sure that everyone starts out marching to the same beat and, doing that, it is the first building block in a successful operation.

Never mind that one venture capitalist, William Congleton, lamented that "Only one thing is certain about a new venture: it's going to turn out very different from its strategic plan." That simply alerts us to the fact that a good strategic plan is flexible. It is enough to know that the absence of a plan, fragile as it is, must be considered unacceptable.

The sixth key in entrepreneurship is composed of two points, *quality* and *cash flow.*

In the beginning, airline start-ups settling on the level of market penetration must focus on reliability and service levels to make sure they are consistent and unflagging.

Having identified the quality of service, it is axiomatic that the level be adhered to without deviation. As to cash flow, start-up efforts are so cash-intensive that entrepreneurs cannot see far beyond the short term. It is not a time to obtain 747 aircraft in anticipation of five-year growth. The entrepreneur cannot afford to enter markets that will not immediately develop, for resources to sustain the operation are not typically available.

The final elements needed for success are the most elusive: luck and timing. Stanley Marcus of Nieman-Marcus said it well: "Never underestimate the value of luck, but remember that luck comes to those searching for something." Or, to invoke a more common clubhouse cliche: "You make your own luck, good or bad." This leads full circle, of course, right back to the entrepreneur himself (or herself). Whether he or she is a person who sees and wants to exploit an obvious opportunity, one who has reached a plateau in a present vocation, or simply one who has a yearning to break out of a mold, it is the personality, intuition, energy, restlessness, and conviction that sets enterprises in motion. The entrepreneur goes to the gaming table with an overwhelming feeling of luck and lets the chips fall where they may.

SUMMARY

The number of successful airline entrepreneurs, either within companies or without, who struck out on their own are so numerous as to support the entrepreneurial ethic.

We may even conclude, therefore, that the need for entrepreneurial innovation is undiminished. As markets grow, niches invariably develop and there will be no shortage of potential opportunities to be exploited. Entrepreneurs are shaking off the 50 years of regulation that beset the aviation industry. Air transportation that served society in the past is quite different today. The impact of present and future trends is still to be felt. The successful entrepreneur must be sensitive to these changes and adapt to them creatively.

Index